BRITISH ASSOCIATION FOR THE ADVANCEMENT OF SCIENCE

MATHEMATICAL TABLES
PART-VOLUME A

LEGENDRE POLYNOMIALS

PREPARED BY THE COMMITTEE
FOR THE
CALCULATION OF MATHEMATICAL TABLES

Published for the British Association

AT THE

UNIVERSITY PRESS
CAMBRIDGE
1946

CAMBRIDGE
UNIVERSITY PRESS

University Printing House, Cambridge CB2 8BS, United Kingdom

Cambridge University Press is part of the University of Cambridge.

It furthers the University's mission by disseminating knowledge in the pursuit of education, learning and research at the highest international levels of excellence.

www.cambridge.org
Information on this title: www.cambridge.org/9781316611944

© Cambridge University Press 1946

First published 1946
First paperback edition 2016

A catalogue record for this publication is available from the British Library

ISBN 978-1-316-61194-4 Paperback

PREFACE

One of the primary functions of the Committee for the Calculation of Mathematical Tables is the production of new tables in response to special demands on the part of research workers. The accompanying tables originated in such a demand but, at the time they were computed, they were thought to be too small for publication as a separate volume in the British Association's series of mathematical tables. It is obviously desirable that new tables for which a definite need has been shown to exist should be made generally available without delay. The Committee has, therefore, decided to begin the publication of a new series of tables, in the form of Part-Volumes, which will include not only tables of the kind indicated, but also other tables which, by reason of their size or nature, can properly be issued therein. It is intended that, in course of time, these Part-Volumes, or such of them as are deemed suitable, shall be combined into volumes and take their place in the Association's main series of tables.

The present tables of Legendre Polynomials, which form the first of the new series, and are designated Part-Volume A, were designed by Dr L. J. Comrie, and were largely computed under his supervision. The Introduction, for which the Committee is also indebted to Dr Comrie, describes the material on which the tables are based, and the methods used in their construction.

For a detailed bibliography of tables of the Legendre functions, including the polynomials and associated functions, reference may be made to the *Index of Mathematical Tables*, by A. Fletcher, J. C. P. Miller and L. Rosenhead, 1946 (London: Scientific Computing Service, Limited).

A. J. THOMPSON

INTRODUCTION

The following tables of Legendre Polynomials, $P_n(x)$, have been prepared to meet the needs of workers in various branches of mathematics and physics.

Tables for $x = 0(0·01)1$. When the accompanying tables were designed (in 1932) the existing tables for this range and interval were:

(1) British Association: Report of the Committee on Mathematical Tables for 1879. This extends to $P_7(x)$, and all values are exact; no differences are given.

(2) Tallqvist, H. Tafeln der Kugelfunctionen $P_n(x)$ und ihrer abgeleiteten Functionen. *Acta Societatis Scientiarum Fennicae*, tome XXXII, no. 6 (Helsingfors, 1904). This includes all the values given in (1) and also $P_8(x)$. All the derivatives are tabulated.

(3) Hayashi, K. *Tafeln der Besselschen, Theta-, Kugel- und anderer Funktionen* (Berlin: Springer, 1930). These tables extend to $P_8(x)$, with no differences or derivatives.

These three tables were compared with each other. No errors were found in (1) and (2) but five occur in Hayashi. He remarks "Die Tafeln der Kugelfunktionen sind aufs neue von mir berechnet. Ein Vergleichen mit den vorhandenen Tafeln für diese Funktionen zeigte ziemlich viele Unstimmigkeiten".

In the present tables, which extend to $n = 12$, $P_2(x)$ and $P_3(x)$ are exact to five and seven decimals respectively; for other values of n, $P_n(x)$ is rounded off to seven decimals. $P_4(x)$ is exact when x is a multiple of 0·02; in the remaining cases it can be made exact by adding 0·4375 units of the seventh decimal; thus

$$P_4(0·01) = +0·37462\ 50437\ 5 \quad \text{and} \quad P_4(0·35) = -0·01872\ 26562\ 5.$$

$P_2(x)$, $P_3(x)$ and $P_4(x)$ were built up from their constant differences on a Burroughs Class 11 machine. $P_5(x)$, $P_6(x)$ and $P_7(x)$ were taken from the 1879 Report of the British Association. $P_8(x)$ is based on Tallqvist and Hayashi, the differences being computed specially for this work. $P_9(x)$ was found to nine decimals by the recurrence formula, and checked by differences. As a further check, $P_8(x)$ and $P_9(x)$ were computed directly at interval 0·2 from the definitions. $P_{10}(x)$, $P_{11}(x)$ and $P_{12}(x)$ were computed by A. J. Thompson from the definitions, without reference to any previous work.

Tables for $x = 1(0·01)6$. These tables, which extend to $n = 12$, are completely new. They give seven significant figures throughout, and usually eight when the first is 1 or 2. Exact values at interval 1·0 were first prepared from the definitions. Then exact values at interval 0·2 were computed as far as $n = 9$, by summation on a Burroughs machine from the constant difference; these values were used later as independent checks. $P_2(x)$, which is exact to five decimals, was built up from its constant second difference of 30; $P_3(x)$ was built up from its second differences, utilising cycles in the end figures; $P_4(x)$, $P_5(x)$ and $P_6(x)$ were built up in the same way, and rounded off as required. The remaining values were computed by the recurrence formula, and checked by differencing on a National machine.*

The printer's copy was prepared on the Burroughs machine already mentioned, by building up from the second, fourth or sixth difference as required. By this means it is possible to ensure the complete accuracy and legibility of every figure.

Tables for $x=6$ (o·1)11. These tables, on p. A 42, are an extension at a larger interval to $x=11$ and $n=6$. The values of $P_2(x)$ and $P_3(x)$ are exact.

Interpolation. Interpolation may be performed by the methods described in the Association's *Mathematical Tables*, vol. 1,† p. vi, or in *Interpolation and Allied Tables* (London: H.M. Stationery Office, 1936). The method described on p. xi of the first-mentioned volume or on pp. 928–9 of the latter volume for eliminating higher-order differences that do not exceed a certain limiting value (1000 for δ^4 and 10,000 for δ^6) has been adopted here; modified differences are denoted by $*\delta^2$ and $*\delta^4$. All the differences necessary for interpolation to the full accuracy of the table are given. Modified differences are to be used in exactly the same way as unmodified differences. Where the modification begins or ends in the middle of a page all the differences provided are to be used; thus an interpolation of $P_7(x)$ for $x=0\cdot795$ would be performed with only one difference (here $+738$) from the column δ^4.

L. J. COMRIE

* Comrie, L. J., "Scientific Applications of the National Accounting Machine." Supplement to the *Journal of the Royal Statistical Society*, **3**, 87 (1936).

† First edition, 1931. Detailed treatment of interpolation is unfortunately omitted from the second edition, 1946.

LEGENDRE POLYNOMIALS

Legendre's differential equation:

$$(1 - x^2)\frac{d^2y}{dx^2} - 2x\frac{dy}{dx} + n(n+1)\,y = 0.$$

Standardised polynomial solution for positive integral values of n:

$$y = P_n(x) = \frac{(2n)!}{2^n(n!)^2}\left[x^n - \frac{n(n-1)}{2(2n-1)}x^{n-2} + \frac{n(n-1)(n-2)(n-3)}{2.4(2n-1)(2n-3)}x^{n-4} - \cdots\right].$$

Explicit expression of polynomials for $n = 0$ to 12:

$$P_0(x) = 1,$$
$$P_1(x) = x,$$
$$P_2(x) = \tfrac{1}{2}(3x^2 - 1),$$
$$P_3(x) = \tfrac{1}{2}(5x^3 - 3x),$$

$$P_4(x) = \tfrac{1}{8}(35x^4 - 30x^2 + 3),$$
$$P_5(x) = \tfrac{1}{8}(63x^5 - 70x^3 + 15x),$$
$$P_6(x) = \tfrac{1}{16}(231x^6 - 315x^4 + 105x^2 - 5),$$
$$P_7(x) = \tfrac{1}{16}(429x^7 - 693x^5 + 315x^3 - 35x),$$

$$P_8(x) = \tfrac{1}{128}(6435x^8 - 12012x^6 + 6930x^4 - 1260x^2 + 35),$$
$$P_9(x) = \tfrac{1}{128}(12155x^9 - 25740x^7 + 18018x^5 - 4620x^3 + 315x),$$
$$P_{10}(x) = \tfrac{1}{256}(46189x^{10} - 109395x^8 + 90090x^6 - 30030x^4 + 3465x^2 - 63),$$
$$P_{11}(x) = \tfrac{1}{256}(88179x^{11} - 230945x^9 + 218790x^7 - 90090x^5 + 15015x^3 - 693x),$$
$$P_{12}(x) = \tfrac{1}{1024}(676039x^{12} - 1939938x^{10} + 2078505x^8 - 1021020x^6 + 225225x^4 - 18018x^2 + 231).$$

Other definitions:

$$(1 - 2xz + z^2)^{-\frac{1}{2}} = P_0(x) + P_1(x)\,z + P_2(x)\,z^2 + P_3(x)\,z^3 + \cdots \quad (z < 1),$$

$$P_n(x) = \frac{1}{2^n n!}\left(\frac{d}{dx}\right)^n (x^2 - 1)^n.$$

Recurrence formula, used (inter alia) to obtain $P_n(x)$ for $n > 12$:

$$(n+1)\,P_{n+1}(x) - (2n+1)\,xP_n(x) + nP_{n-1}(x) = 0.$$

Derivative formulae:

$$(x^2 - 1)\,P_n'(x) = n\{xP_n(x) - P_{n-1}(x)\} = (n+1)\{P_{n+1}(x) - xP_n(x)\},$$

$$P_n'(x) = (2n-1)\,P_{n-1}(x) + (2n-5)\,P_{n-3}(x) + (2n-9)\,P_{n-5}(x) + \cdots,$$

$$xP_n'(x) - P_{n-1}'(x) = nP_n(x), \qquad\qquad P_{n+1}'(x) - xP_n'(x) = (n+1)\,P_n(x),$$

$$P_{n+1}'(x) - P_{n-1}'(x) = (2n+1)\,P_n(x), \qquad P_{n+1}'(x) - 2xP_n'(x) + P_{n-1}'(x) = P_n(x).$$

Special values:

$$P_{2n}(0) = (-1)^n\frac{(2n)!}{2^{2n}(n!)^2}, \qquad P_{2n+1}(0) = 0, \qquad P_n(1) = 1, \qquad P_n(-x) = (-1)^n\,P_n(x).$$

Orthogonal relation:

$$\int_{-1}^{1} P_m(x)\,P_n(x)\,dx = 0,\ \text{when } m \neq n; \qquad \int_{-1}^{1}\{P_n(x)\}^2\,dx = \frac{2}{2n+1}.$$

Expression of x^{n-1} in terms of Legendre polynomials:

$$x^{n-1} = \frac{2^{n-1}\{(n-1)!\}^2}{(2n-1)!}\left[(2n-1)\,P_{n-1}(x) + (2n-5)\frac{(2n-1)}{2}P_{n-3}(x)\right.$$
$$\left. + (2n-9)\frac{(2n-1)(2n-3)}{2.4}P_{n-5}(x) + (2n-13)\frac{(2n-1)(2n-3)(2n-5)}{2.4.6}P_{n-7}(x) + \cdots\right].$$

Illustrations of accuracy in use of recurrence formula:

(i) $11P_{11}(\cdot73) = 21(\cdot73)\,P_{10}(\cdot73) - 10P_9(\cdot73) = 21(\cdot73)\,(\cdot20299\ 76) - 10(\cdot31199\ 70) = -\cdot00801\ 68$,
hence $P_{11}(\cdot73) = -\cdot00072\ 88\ (-\cdot00072\ 87$ in the tables);

(ii) $12P_{12}(\cdot73) = 23(\cdot73)\,P_{11}(\cdot73) - 11P_{10}(\cdot73) = 23(\cdot73)\,(-\cdot00072\ 87) - 11(\cdot20299\ 76) = -2\cdot24520\ 85$,
hence $P_{12}(\cdot73) = -\cdot18710\ 07\ (-\cdot18710\ 08$ in the tables).

Everett's interpolation formula $(\phi = 1 - \theta)$:

$$u_\theta = \phi u_0 + \frac{\phi(\phi^2 - 1)}{3!}\delta^2 u_0 + \frac{\phi(\phi^2 - 1)(\phi^2 - 4)}{5!}\delta^4 u_0 + \cdots + \theta u_1 + \frac{\theta(\theta^2 - 1)}{3!}\delta^2 u_1 + \frac{\theta(\theta^2 - 1)(\theta^2 - 4)}{5!}\delta^4 u_1 + \cdots.$$

x	$P_2(x)$†	$P_3(x)$	δ^2	$P_4(x)$	δ^2	$P_5(x)$	*δ^2	$P_6(x)$	*δ^2
0·00	− 0·50000	− 0·00000 00	0	+ 0·37500 00	− 75 00	+ 0·00000 00	0	− 0·31250 00	+ 131 31
·01	·49985	·01499 75	+ 1 50	·37462 50	74 93	·01874 13	− 5 27	·31184 39	131 05
·02	·49940	·02998 00	3 00	·37335 07	74 79	·03743 00	10 48	·30987 81	130 34
·03	·49865	·04493 25	4 50	·37162 85	74 51	·05601 39	15 70	·30660 97	129 19
·04	·49760	·05984 00	6 00	·36901 12	74 16	·07444 08	20 91	·30205 03	127 51
0·05	− 0·49625	− 0·07468 75	+ 7 50	+ 0·36565 23	− 73 67	+ 0·09265 87	− 26 06	− 0·29621 66	+ 125 43
·06	·49460	·08946 00	9 00	·36155 67	73 11	·11061 61	31 16	·28912 95	122 86
·07	·49265	·10414 25	10 50	·35673 00	72 41	·12826 20	36 23	·28081 47	119 81
·08	·49040	·11872 00	12 00	·35117 92	71 64	·14554 58	41 19	·27130 26	116 35
·09	·48785	·13317 75	13 50	·34491 20	70 73	·16241 78	46 12	·26062 78	112 45
0·10	− 0·48500	− 0·14750 00	+ 15 00	+ 0·33793 75	− 69 75	+ 0·17882 88	− 50 94	− 0·24882 93	+ 108 10
·11	·48185	·16167 25	16 50	·33026 55	68 63	·19473 06	55 65	·23595 06	103 33
·12	·47840	·17568 00	18 00	·32190 72	67 44	·21007 60	60 31	·22203 93	98 18
·13	·47465	·18950 75	19 50	·31287 45	66 11	·22481 86	64 79	·20714 70	92 60
·14	·47060	·20314 00	21 00	·30318 07	64 71	·23891 35	69 18	·19132 94	86 63
0·15	− 0·46625	− 0·21656 25	+ 22 50	+ 0·29283 98	− 63 17	+ 0·25231 68	− 73 46	− 0·17464 61	+ 80 33
·16	·46160	·22976 00	24 00	·28186 72	61 56	·26498 58	77 57	·15716 02	73 64
·17	·45665	·24271 75	25 50	·27027 90	59 81	·27687 94	81 53	·13893 85	66 63
·18	·45140	·25542 00	27 00	·25809 27	57 99	·28795 80	85 32	·12005 11	59 27
·19	·44585	·26785 25	28 50	·24532 65	56 03	·29818 37	88 98	·10057 15	51 65
0·20	− 0·44000	− 0·28000 00	+ 30 00	+ 0·23200 00	− 54 00	+ 0·30752 00	− 92 41	− 0·08057 60	+ 43 71
·21	·43385	·29184 75	31 50	·21813 35	51 83	·31593 25	95 69	·06014 39	35 50
·22	·42740	·30338 00	33 00	·20374 87	49 59	·32338 85	98 75	·03935 72	27 07
·23	·42065	·31458 25	34 50	·18886 80	47 21	·32985 74	101 61	− ·01830 02	18 42
·24	·41360	·32544 00	36 00	·17351 52	44 76	·33551 06	104 25	+ ·00294 06	9 56
0·25	− 0·40625	− 0·33593 75	+ 37 50	+ 0·15771 48	− 42 17	+ 0·33972 17	− 106 66	+ 0·02427 67	+ 53
·26	·39860	·34606 00	39 00	·14149 27	39 51	·34306 66	108 85	·04561 78	− 8 65
·27	·39065	·35579 25	40 50	·12487 55	36 71	·34532 35	110 76	·06687 22	17 96
·28	·38240	·36512 00	42 00	·10789 12	33 84	·34647 32	112 46	·08794 69	27 34
·29	·37385	·37402 75	43 50	·09056 85	30 83	·34649 88	113 87	·10874 81	36 81
0·30	− 0·36500	− 0·38250 00	+ 45 00	+ 0·07293 75	− 27 75	+ 0·34538 62	− 114 99	+ 0·12918 12	− 46 29
·31	·35585	·39052 25	46 50	·05502 90	24 53	·34312 42	115 87	·14915 14	55 80
·32	·34640	·39808 00	48 00	·03687 52	21 24	·33970 41	116 41	·16856 37	65 25
·33	·33665	·40515 75	49 50	+ ·01850 90	17 81	·33512 04	116 69	·18732 36	74 67
·34	·32660	·41174 00	51 00	− ·00003 53	14 31	·32937 04	116 63	·20533 70	84 00
0·35	− 0·31625	− 0·41781 25	+ 52 50	− 0·01872 27	− 10 67	+ 0·32245 47	− 116 25	+ 0·22251 07	− 93 18
·36	·30560	·42336 00	54 00	·03751 68	6 96	·31437 71	115 55	·23875 29	102 19
·37	·29465	·42836 75	55 50	·05638 05	− 3 11	·30514 46	114 50	·25397 35	111 04
·38	·28340	·43282 00	57 00	·07527 53	+ 81	·29476 77	113 12	·26808 42	119 60
·39	·27185	·43670 25	58 50	·09416 20	4 87	·28326 03	111 36	·28099 94	127 92
0·40	− 0·26000	− 0·44000 00	+ 60 00	− 0·11300 00	+ 9 00	+ 0·27064 00	− 109 24	+ 0·29263 60	− 135 90
·41	·24785	·44269 75	61 50	·13174 80	13 27	·25692 80	106 74	·30291 42	143 55
·42	·23540	·44478 00	63 00	·15036 33	17 61	·24214 93	103 84	·31175 77	150 74
·43	·22265	·44623 25	64 50	·16880 25	22 09	·22633 29	100 58	·31909 45	157 57
·44	·20960	·44704 00	66 00	·18702 08	26 64	·20951 15	96 88	·32485 66	163 83
0·45	− 0·19625	− 0·44718 75	+ 67 50	− 0·20497 27	+ 31 33	+ 0·19172 21	− 92 77	+ 0·32898 13	− 169 61
·46	·18260	·44666 00	69 00	·22261 13	36 09	·17300 58	88 24	·33141 10	174 78
·47	·16865	·44544 25	70 50	·23988 90	40 99	·15340 79	83 27	·33209 40	179 36
·48	·15440	·44352 00	72 00	·25675 68	45 96	·13297 81	77 86	·33098 47	183 21
·49	·13985	·44087 75	73 50	·27316 50	51 07	·11177 05	71 99	·32804 46	186 37
0·50	− 0·12500	− 0·43750 00	+ 75 00	− 0·28906 25	+ 56 25	+ 0·08984 38	− 65 70	+ 0·32324 22	− 188 77

* These differences have been modified; see page A 4. † For $P_2(x)$, δ^2 is + 30 throughout.

LEGENDRE POLYNOMIALS

x	$P_2(x)$†	$P_3(x)$	δ^2	$P_4(x)$	δ^2	$P_5(x)$	*δ^2	$P_6(x)$	*δ^2
0·50	− 0·12500	− 0·43750 00	+ 75 00	− 0·28906 25	+ 56 25	+ 0·08984 38	− 65 70	+ 0·32324 22	− 188 77
·51	·10985	·43337 25	76 50	·30439 75	61 57	·06726 11	58 85	·31655 37	190 30
·52	·09440	·42848 00	78 00	·31911 68	66 96	·04409 07	51 61	·30796 38	190 98
·53	·07865	·42280 75	79 50	·33316 65	72 49	+ ·02040 52	43 81	·29746 58	190 72
·54	·06260	·41634 00	81 00	·34649 13	78 09	− ·00371 75	35 56	·28506 24	189 47
0·55	− 0·04625	− 0·40906 25	+ 82 50	− 0·35903 52	+ 83 83	− 0·02819 48	− 26 75	+ 0·27076 62	− 187 18
·56	·02960	·40096 00	84 00	·37074 08	89 64	·05293 87	17 47	·25460 02	183 78
·57	− ·01265	·39201 75	85 50	·38155 00	95 59	·07785 63	− 7 60	·23659 85	179 25
·58	+ ·00460	·38222 00	87 00	·39140 33	101 61	·10284 90	+ 2 72	·21680 66	173 47
·59	·02215	·37155 25	88 50	·40024 05	107 77	·12781 34	13 69	·19528 23	166 45
0·60	+ 0·04000	− 0·36000 00	+ 90 00	− 0·40800 00	+ 114 00	− 0·15264 00	+ 25 13	+ 0·17209 60	− 158 06
·61	·05815	·34754 75	91 50	·41461 95	120 37	·17721 42	37 20	·14733 17	148 27
·62	·07660	·33418 00	93 00	·42003 53	126 81	·20141 54	49 80	·12108 73	137 07
·63	·09535	·31988 25	94 50	·42418 30	133 39	·22511 75	63 03	·09347 51	124 28
·64	·11440	·30464 00	96 00	·42699 68	140 04	·24818 83	76 80	·06462 30	109 92
0·65	+ 0·13375	− 0·28843 75	+ 97 50	− 0·42841 02	+ 146 83	− 0·27048 99	+ 91 23	+ 0·03467 47	− 93 93
·66	·15340	·27126 00	99 00	·42835 53	153 69	·29187 81	106 25	+ ·00379 03	76 20
·67	·17335	·25309 25	100 50	·42676 35	160 69	·31220 27	121 88	− ·02785 28	56 66
·68	·19360	·23392 00	102 00	·42356 48	167 76	·33130 73	138 16	·06005 91	35 29
·69	·21415	·21372 75	103 50	·41868 85	174 97	·34902 91	155 09	·09261 47	− 11 97
0·70	+ 0·23500	− 0·19250 00	+ 105 00	− 0·41206 25	+ 182 25	− 0·36519 88	+ 172 67	− 0·12528 63	+ 13 35
·71	·25615	·17022 25	106 50	·40361 40	189 67	·37964 06	190 87	·15782 06	40 75
·72	·27760	·14688 00	108 00	·39326 88	197 16	·39217 24	209 81	·18994 35	70 26
·73	·29935	·12245 75	109 50	·38095 20	204 79	·40260 49	229 38	·22135 96	102 07
·74	·32140	·09694 00	111 00	·36658 73	212 49	·41074 23	249 66	·25175 09	136 07
0·75	+ 0·34375	− 0·07031 25	+ 112 50	− 0·35009 77	+ 220 33	− 0·41638 18	+ 270 61	− 0·28077 70	+ 172 54
·76	·36640	·04256 00	114 00	·33140 48	228 24	·41931 38	292 34	·30807 32	211 41
·77	·38935	− ·01366 75	115 50	·31042 95	236 29	·41932 12	314 70	·33325 06	252 83
·78	·41260	+ ·01638 00	117 00	·28709 13	244 41	·41618 02	337 86	·35589 49	296 84
·79	·43615	·04759 75	118 50	·26130 90	252 67	·40965 93	361 70	·37556 58	343 56
0·80	+ 0·46000	+ 0·08000 00	+ 120 00	− 0·23300 00	+ 261 00	− 0·39952 00	+ 386 34	− 0·39179 60	+ 393 05
·81	·48415	·11360 25	121 50	·20208 10	269 47	·38551 60	411 69	·40409 05	445 38
·82	·50860	·14842 00	123 00	·16846 73	278 01	·36739 37	437 83	·41192 58	500 64
·83	·53335	·18446 75	124 50	·13207 35	286 69	·34489 17	464 72	·41474 91	558 96
·84	·55840	·22176 00	126 00	·09281 28	295 44	·31774 10	492 45	·41197 71	620 35
0·85	+ 0·58375	+ 0·26031 25	+ 127 50	− 0·05059 77	+ 304 33	− 0·28566 44	+ 520 91	− 0·40299 57	+ 684 96
·86	·60940	·30014 00	129 00	− ·00533 93	313 29	·24837 72	550 20	·38715 87	752 83
·87	·63535	·34125 75	130 50	+ ·04305 20	322 39	·20558 65	580 31	·36378 72	824 10
·88	·66160	·38368 00	132 00	·09466 72	331 56	·15699 12	611 25	·33216 84	898 79
·89	·68815	·42742 25	133 50	·14959 80	340 87	·10228 19	642 98	·29155 51	977 09
0·90	+ 0·71500	+ 0·47250 00	+ 135 00	+ 0·20793 75	+ 350 25	− 0·04114 12	+ 675 58	− 0·24116 43	+ 1058 97
·91	·74215	·51892 75	136 50	·26977 95	359 77	+ ·02675 69	709 04	·18017 69	1144 64
·92	·76960	·56672 00	138 00	·33521 92	369 36	·10174 70	743 35	·10773 61	1234 12
·93	·79735	·61589 25	139 50	·40435 25	379 09	·18417 22	778 52	− ·02294 69	1327 50
·94	·82540	·66646 00	141 00	·47727 67	388 89	·27438 42	814 57	+ ·07512 48	1424 95
0·95	+ 0·85375	+ 0·71843 75	+ 142 50	+ 0·55408 98	+ 398 83	+ 0·37274 36	+ 851 54	+ 0·18745 36	+ 1526 51
·96	·88240	·77184 00	144 00	·63489 12	408 84	·47962 00	889 36	·31505 52	1632 27
·97	·91135	·82668 25	145 50	·71978 10	418 99	·59539 17	928 12	·45898 74	1742 34
·98	·94060	·88298 00	147 00	·80886 07	429 21	·72044 63	967 78	·62035 12	1856 88
0·99	0·97015	0·94074 75	148 50	0·90223 25	439 57	0·85518 04	1008 38	0·80029 20	1975 86
1·00	+ 1·00000	+ 1·00000 00	+ 150 00	+ 1·00000 00	+ 450 00	+ 1·00000 00	+ 1049 91	+ 1·00000 00	+ 2099 54

* These differences have been modified; see page A 4. † For $P_2(x)$, δ^2 is + 30 throughout.

x	$P_2(x)$†	$P_3(x)$	δ^2	$P_4(x)$	δ^2	$P_5(x)$	*δ^2	$P_6(x)$	*δ^2
1·00	1·00000	1·00000 00	150 00	1·00000 00	450 00	1·00000 00	1049 91	1·00000 00	2099 54
·01	1·03015	1·06075 25	151 50	1·10226 75	460 57	1·15532 04	1092 37	1·22071 20	2227 89
·02	1·06060	1·12302 00	153 00	1·20914 07	471 21	1·32156 63	1135 82	1·46371 18	2361 09
·03	1·09135	1·18681 75	154 50	1·32072 60	481 99	1·49917 21	1180 19	1·73033 16	2499 22
·04	1·12240	1·25216 00	156 00	1·43713 12	492 84	1·68858 16	1225 58	2·02195 29	2642 42
1·05	1·15375	1·31906 25	157 50	1·55846 48	503 83	1·89024 86	1271 89	2·34000 78	2790 70
·06	1·18540	1·38754 00	159 00	1·68483 67	514 89	2·10463 64	1319 26	2·68597 95	2944 33
·07	1·21735	1·45776 75	160 50	1·81635 75	526 09	2·33221 86	1367 59	3·06140 43	3103 23
·08	1·24960	1·52928 00	162 00	1·95313 92	537 36	2·57347 86	1416 94	3·46787 16	3267 71
·09	1·28215	1·60257 25	163 50	2·09529 45	548 77	2·82890 99	1467 31	3·90702 62	3437 68
1·10	1·31500	1·67750 00	165 00	2·24293 75	560 25	3·09901 62	1518 73	4·38056 8	3613 4
·11	1·34815	1·75407 75	166 50	2·39618 30	571 87	3·38431 17	1571 17	4·89025 5	3795 1
·12	1·38160	1·83232 00	168 00	2·55514 72	583 56	3·68532 08	1624 64	5·43790 3	3982 2
·13	1·41535	1·91224 25	169 50	2·71994 70	595 39	4·00257 83	1679 20	6·02538 5	4176 1
·14	1·44940	1·99386 00	171 00	2·89070 07	607 29	4·33662 98	1734 83	6·65463 9	4375 6
1·15	1·48375	2·07718 75	172 50	3·06752 73	619 33	4·68803 16	1791 52	7·32766 1	4581 5
·16	1·51840	2·16224 00	174 00	3·25054 72	631 44	5·05735 06	1849 28	8·04651 0	4793 9
·17	1·55335	2·24903 25	175 50	3·43988 15	643 69	5·44516 45	1908 19	8·81331 0	5012 9
·18	1·58860	2·33758 00	177 00	3·63565 27	656 01	5·85206 23	1968 16	9·63025 1	5238 3
·19	1·62415	2·42789 75	178 50	3·83798 40	668 47	6·27864 38	2029 26	10·49958 8	5470 6
1·20	1·66000	2·52000 00	180 00	4·04700 00	681 00	6·72552 0	2091 5	11·42364 4	5709 8
·21	1·69615	2·61390 25	181 50	4·26282 60	693 67	7·19331 3	2154 9	12·40481 1	5956 1
·22	1·73260	2·70962 00	183 00	4·48558 87	706 41	7·68265 7	2219 3	13·44555 2	6209 2
·23	1·76935	2·80716 75	184 50	4·71541 55	719 29	8·19419 6	2285 0	14·54839 9	6470 0
·24	1·80640	2·90656 00	186 00	4·95243 52	732 24	8·72858 7	2351 9	15·71595 9	6737 9
1·25	1·84375	3·00781 25	187 50	5·19677 73	745 33	9·28649 9	2419 8	16·95091 2	7013 5
·26	1·88140	3·11094 00	189 00	5·44857 27	758 49	9·86861 1	2489 0	18·25601 4	7296 4
·27	1·91935	3·21595 75	190 50	5·70795 30	771 79	10·47561 5	2559 3	19·63409 5	7587 6
·28	1·95760	3·32288 00	192 00	5·97505 12	785 16	11·10821 4	2630 9	21·08806 6	7886 3
·29	1·99615	3·43172 25	193 50	6·25000 10	798 67	11·76712 4	2703 8	22·62091 5	8193 2
1·30	2·03500	3·54250 00	195 00	6·53293 8	812 2	12·45307 4	2777 5	24·23571 1	8508 3
·31	2·07415	3·65522 75	196 50	6·82399 7	825 9	13·16680 2	2852 9	25·93560 5	8831 6
·32	2·11360	3·76992 00	198 00	7·12331 5	839 9	13·90906 1	2929 4	27·72383 1	9163 6
·33	2·15335	3·88659 25	199 50	7·43103 2	853 6	14·68061 6	3006 9	29·60370 8	9503 8
·34	2·19340	4·00526 00	201 00	7·74728 5	867 7	15·48224 3	3086 0	31·57863 9	9853 0
1·35	2·23375	4·12593 75	202 50	8·07221 5	881 8	16·31473 2	3166 3	33·65211 6	10211 1
·36	2·27440	4·24864 00	204 00	8·40596 3	896 1	17·17888 6	3247 6	35·82772 0	10577 7
·37	2·31535	4·37338 25	205 50	8·74867 2	910 4	18·07551 9	3330 6	38·10911 9	10954 3
·38	2·35660	4·50018 00	207 00	9·10048 5	924 8	19·00546 0	3414 5	40·50007 7	11339 3
·39	2·39815	4·62904 75	208 50	9·46154 6	939 3	19·96954 9	3500 0	43·00444 6	11734 4
1·40	2·44000	4·76000 00	210 00	9·83200 0	954 1	20·96864 0	3586 6	45·62617 6	12138 7
·41	2·48215	4·89305 25	211 50	10·21199 5	968 7	22·00360 0	3674 8	48·36931 1	12552 8
·42	2·52460	5·02822 00	213 00	10·60167 7	983 6	23·07531 0	3763 9	51·23799 2	12976 9
·43	2·56735	5·16551 75	214 50	11·00119 5	998 6	24·18466 2	3854 8	54·23646 0	13410 7
·44	2·61040	5·30496 00	216 00	11·41069 9	1013 7	25·33256 4	3946 9	57·36905 4	13854 9
1·45	2·65375	5·44656 25	217 50	11·83034 0	1028 8	26·51993 7	4040 1	60·64021 6	14309 4
·46	2·69740	5·59034 00	219 00	12·26026 9	1044 1	27·74771 4	4135 1	64·05449 1	14774 2
·47	2·74135	5·73630 75	220 50	12·70063 9	1059 4	29·01684 4	4231 0	67·61652 8	15249 9
·48	2·78560	5·88448 00	222 00	13·15160 3	1075 1	30·32828 7	4328 8	71·33108 3	15735 8
·49	2·83015	6·03487 25	223 50	13·61331 8	1090 5	31·68302 0	4427 5	75·20301 7	16233 5
1·50	2·87500	6·18750 00	225 00	14·08593 8	1106 2	33·08203 1	4528 1	79·23730 5	16741 2

* These differences have been modified; see page A 4. † For $P_2(x)$, δ^2 is $+30$ throughout.

(A 8)

LEGENDRE POLYNOMIALS

x	$P_2(x)$†	$P_3(x)$	δ^2	$P_4(x)$	δ^2	$P_5(x)$	δ^2	$P_6(x)$	*δ^2
1·50	2·87500	6·18750 0	225 0	14·08593 8	1106 2	33·08203	4529	79·23730	16744
·51	2·92015	6·34237 8	226 4	14·56962 0	1122 1	34·52633	4629	83·43903	17259
·52	2·96560	6·49952 0	228 0	15·06452 3	1138 0	36·01692	4734	87·81338	17792
·53	3·01135	6·65894 2	229 6	15·57080 6	1154 0	37·55485	4837	92·36567	18335
·54	3·05740	6·82066 0	231 0	16·08862 9	1170 0	39·14115	4945	97·10133	18886
1·55	3·10375	6·98468 8	232 4	16·61815 2	1186 4	40·77690	5050	102·02588	19455
·56	3·15040	7·15104 0	234 0	17·15953 9	1202 7	42·46315	5162	107·14500	20033
·57	3·19735	7·31973 2	235 6	17·71295 3	1219 0	44·20102	5270	112·46447	20621
·58	3·24460	7·49078 0	237 0	18·27855 7	1235 6	45·99159	5383	117·99018	21225
·59	3·29215	7·66419 8	238 4	18·85651 7	1252 3	47·83599	5497	123·72816	21840
1·60	3·34000	7·84000 0	240 0	19·44700 0	1269 0	49·73536	5611	129·68456	22466
·61	3·38815	8·01820 2	241 6	20·05017 3	1285 9	51·69084	5728	135·86565	23109
·62	3·43660	8·19882 0	243 0	20·66620 5	1302 8	53·70360	5845	142·27785	23760
·63	3·48535	8·38186 8	244 4	21·29526 5	1319 8	55·77481	5966	148·92768	24427
·64	3·53440	8·56736 0	246 0	21·93752 3	1337 1	57·90568	6086	155·82181	25109
1·65	3·58375	8·75531 2	247 6	22·59315 2	1354 4	60·09741	6209	162·96705	25800
·66	3·63340	8·94574 0	249 0	23·26232 5	1371 6	62·35123	6334	170·37032	26509
·67	3·68335	9·13865 8	250 4	23·94521 4	1389 2	64·66839	6458	178·03870	27229
·68	3·73360	9·33408 0	252 0	24·64199 5	1406 8	67·05013	6586	185·97940	27964
·69	3·78415	9·53202 2	253 6	25·35284 4	1424 5	69·49773	6716	194·19977	28715
1·70	3·83500	9·73250 0	255 0	26·07793 8	1442 2	72·01249	6845	202·70731	29476
·71	3·88615	9·93552 8	256 4	26·81745 4	1460 1	74·59570	6978	211·50964	30255
·72	3·93760	10·14112 0	258 0	27·57157 1	1478 3	77·24869	7111	220·61455	31050
·73	3·98935	10·34929 2	259 6	28·34047 1	1496 2	79·97279	7247	230·02998	31855
·74	4·04140	10·56006 0	261 0	29·12433 3	1514 5	82·76936	7384	239·76399	32678
1·75	4·09375	10·77343 8	262 4	29·92334 0	1532 8	85·63977	7522	249·82481	33519
·76	4·14640	10·98944 0	264 0	30·73767 5	1551 3	88·58540	7663	260·22084	34370
·77	4·19935	11·20808 2	265 6	31·56752 3	1569 8	91·60766	7805	270·96060	35239
·78	4·25260	11·42938 0	267 0	32·41306 9	1588 4	94·70797	7948	282·05278	36125
·79	4·30615	11·65334 8	268 4	33·27449 9	1607 1	97·88776	8093	293·50624	37025
1·80	4·36000	11·88000 0	270 0	34·15200 0	1626 1	101·14848	8241	305·32998	37945
·81	4·41415	12·10935 2	271 6	35·04576 2	1644 9	104·49161	8389	317·53319	38876
·82	4·46860	12·34142 0	273 0	35·95597 3	1664 0	107·91863	8540	330·12519	39826
·83	4·52335	12·57621 8	274 4	36·88282 4	1683 2	111·43105	8691	343·11548	40794
·84	4·57840	12·81376 0	276 0	37·82650 7	1702 5	115·03038	8847	356·51374	41777
1·85	4·63375	13·05406 2	277 6	38·78721 5	1721 8	118·71818	9000	370·32980	42778
·86	4·68940	13·29714 0	279 0	39·76514 1	1741 3	122·49598	9159	384·57367	43797
·87	4·74535	13·54300 8	280 4	40·76048 0	1760 8	126·36537	9317	399·25554	44832
·88	4·80160	13·79168 0	282 0	41·77342 7	1780 7	130·32793	9479	414·38576	45885
·89	4·85815	14·04317 2	283 6	42·80418 1	1800 3	134·38528	9642	429·97486	46957
1·90	4·91500	14·29750 0	285 0	43·85293 8	1820 2	138·53905	9804	446·03356	48047
·91	4·97215	14·55467 8	286 4	44·91989 7	1840 3	142·79086	9973	462·57276	49153
·92	5·02960	14·81472 0	288 0	46·00525 9	1860 4	147·14240	10139	479·60353	50281
·93	5·08735	15·07764 2	289 6	47·10922 5	1880 6	151·59533	10310	497·13714	51423
·94	5·14540	15·34346 0	291 0	48·23199 7	1900 8	156·15136	10482	515·18502	52590
1·95	5·20375	15·61218 8	292 4	49·37377 7	1921 4	160·81221	10654	533·75883	53771
·96	5·26240	15·88384 0	294 0	50·53477 1	1941 9	165·57960	10830	552·87039	54973
·97	5·32135	16·15843 2	295 6	51·71518 4	1962 4	170·45529	11008	572·53172	56196
·98	5·38060	16·43598 0	297 0	52·91522 1	1983 2	175·44106	11186	592·75504	57436
1·99	5·44015	16·71649 8	298 4	54·13509 0	2004 1	180·53869	11368	613·55276	58699
2·00	5·50000	17·00000 0	300 0	55·37500 0	2025 0	185·75000	11550	634·93750	59979

* These differences have been modified; see page A 4.   † For $P_2(x)$, δ^2 is $+30$ throughout.

LEGENDRE POLYNOMIALS

x	$P_2(x)$†	$P_3(x)$	δ^2	$P_4(x)$	δ^2	$P_5(x)$	δ^2	$P_6(x)$	δ^2
2·00	5·50000	17·00000 0	300 0	55·37500	2025	185·75000	11550	634·9375	5998
·01	5·56015	17·28650 2	301 6	56·63516	2046	191·07681	11734	656·9221	6128
·02	5·62060	17·57602 0	303 0	57·91578	2067	196·52096	11922	679·5195	6261
·03	5·68135	17·86856 8	304 4	59·21707	2089	202·08433	12110	702·7430	6394
·04	5·74240	18·16416 0	306 0	60·53925	2110	207·76880	12301	726·6059	6533
2·05	5·80375	18·46281 2	307 6	61·88253	2131	213·57628	12492	751·1221	6669
·06	5·86540	18·76454 0	309 0	63·24712	2153	219·50868	12687	776·3052	6810
·07	5·92735	19·06935 8	310 4	64·63324	2174	225·56795	12883	802·1693	6954
·08	5·98960	19·37728 0	312 0	66·04110	2197	231·75605	13082	828·7288	7099
·09	6·05215	19·68832 2	313 6	67·47093	2218	238·07497	13281	855·9982	7244
2·10	6·11500	20·00250 0	315 0	68·92294	2240	244·52670	13485	883·9920	7396
·11	6·17815	20·31982 8	316 4	70·39735	2263	251·11328	13687	912·7254	7546
·12	6·24160	20·64032 0	318 0	71·89439	2284	257·83673	13894	942·2134	7701
·13	6·30535	20·96399 2	319 6	73·41427	2307	264·69912	14102	972·4715	7856
·14	6·36940	21·29086 0	321 0	74·95722	2329	271·70253	14312	1003·5152	8016
2·15	6·43375	21·62093 8	322 4	76·52346	2353	278·84906	14525	1035·3605	8176
·16	6·49840	21·95424 0	324 0	78·11323	2373	286·14084	14738	1068·0234	8338
·17	6·56335	22·29078 2	325 6	79·72673	2398	293·58000	14955	1101·5201	8506
·18	6·62860	22·63058 0	327 0	81·36421	2420	301·16871	15173	1135·8674	8673
·19	6·69415	22·97364 8	328 4	83·02589	2443	308·90915	15393	1171·0820	8843
2·20	6·76000	23·32000 0	330 0	84·71200	2466	316·80352	15616	1207·1809	9016
·21	6·82615	23·66965 2	331 6	86·42277	2489	324·85405	15841	1244·1814	9191
·22	6·89260	24·02262 0	333 0	88·15843	2512	333·06299	16066	1282·1010	9369
·23	6·95935	24·37891 8	334 4	89·91921	2537	341·43259	16295	1320·9575	9551
·24	7·02640	24·73856 0	336 0	91·70536	2558	349·96514	16527	1360·7691	9731
2·25	7·09375	25·10156 2	337 6	93·51709	2583	358·66296	16760	1401·5538	9919
·26	7·16140	25·46794 0	339 0	95·35465	2607	367·52838	16993	1443·3304	1·0105
·27	7·22935	25·83770 8	340 4	97·21828	2630	376·56373	17231	1486·1175	1·0297
·28	7·29760	26·21088 0	342 0	99·10821	2654	385·77139	17472	1529·9343	1·0489
·29	7·36615	26·58747 2	343 6	101·02468	2679	395·15377	17711	1574·8000	1·0686
2·30	7·43500	26·96750 0	345 0	102·96794	2701	404·71326	17956	1620·7343	1·0884
·31	7·50415	27·35097 8	346 4	104·93821	2728	414·45231	18202	1667·7570	1·1086
·32	7·57360	27·73792 0	348 0	106·93576	2749	424·37338	18449	1715·8883	1·1289
·33	7·64335	28·12834 2	349 6	108·96080	2776	434·47894	18699	1765·1485	1·1497
·34	7·71340	28·52226 0	351 0	111·01360	2800	444·77149	18953	1815·5584	1·1705
2·35	7·78375	28·91968 8	352 4	113·09440	2824	455·25357	19206	1867·1388	1·1918
·36	7·85440	29·32064 0	354 0	115·20344	2849	465·92771	19464	1919·9110	1·2134
·37	7·92535	29·72513 2	355 6	117·34097	2874	476·79649	19722	1973·8966	1·2351
·38	7·99660	30·13318 0	357 0	119·50724	2899	487·86249	19984	2029·1173	1·2572
·39	8·06815	30·54479 8	358 4	121·70250	2924	499·12833	20247	2085·5952	1·2796
2·40	8·14000	30·96000 0	360 0	123·92700	2949	510·59664	20513	2143·3527	1·3023
·41	8·21215	31·37880 2	361 6	126·18099	2974	522·27008	20781	2202·4125	1·3252
·42	8·28460	31·80122 0	363 0	128·46472	2999	534·15133	21051	2262·7975	1·3484
·43	8·35735	32·22726 8	364 4	130·77844	3026	546·24309	21325	2324·5309	1·3722
·44	8·43040	32·65696 0	366 0	133·12242	3050	558·54810	21598	2387·6365	1·3958
2·45	8·50375	33·09031 2	367 6	135·49690	3077	571·06909	21876	2452·1379	1·4201
·46	8·57740	33·52734 0	369 0	137·90215	3102	583·80884	22156	2518·0594	1·4446
·47	8·65135	33·96805 8	370 4	140·33842	3127	596·77015	22438	2585·4255	1·4693
·48	8·72560	34·41248 0	372 0	142·80596	3155	609·95584	22721	2654·2609	1·4944
·49	8·80015	34·86062 2	373 6	145·30505	3180	623·36874	23008	2724·5907	1·5199
2·50	8·87500	35·31250 0	375 0	147·83594	3206	637·01172	23297	2796·4404	1·5456

† For $P_2(x)$, δ^2 is $+30$ throughout.

x	$P_2(x)$†	$P_3(x)$	δ^2	$P_4(x)$	δ^2	$P_5(x)$	δ^2	$P_6(x)$	δ^2
2·50	8·87500	35·31250 0	375 0	147·83594	3206	637·0117	2330	2796·440	1·546
·51	8·95015	35·76812 8	376 4	150·39889	3232	650·8877	2358	2869·836	1·571
·52	9·02560	36·22752 0	378 0	152·99416	3260	664·9995	2389	2944·803	1·597
·53	9·10135	36·69069 2	379 6	155·62203	3285	679·3502	2417	3021·367	1·626
·54	9·17740	37·15766 0	381 0	158·28275	3312	693·9426	2448	3099·557	1·651
2·55	9·25375	37·62843 8	382 4	160·97659	3339	708·7798	2478	3179·398	1·680
·56	9·33040	38·10304 0	384 0	163·70382	3366	723·8648	2507	3260·919	1·706
·57	9·40735	38·58148 2	385 6	166·46471	3392	739·2005	2540	3344·146	1·735
·58	9·48460	39·06378 0	387 0	169·25952	3419	754·7902	2568	3429·108	1·763
·59	9·56215	39·54994 8	388 4	172·08852	3448	770·6367	2602	3515·833	1·792
2·60	9·64000	40·04000 0	390 0	174·95200	3474	786·7434	2631	3604·350	1·821
·61	9·71815	40·53395 2	391 6	177·85022	3500	803·1132	2663	3694·688	1·850
·62	9·79660	41·03182 0	393 0	180·78344	3530	819·7493	2695	3786·876	1·880
·63	9·87535	41·53361 8	394 4	183·75196	3556	836·6549	2727	3880·944	1·911
·64	9·95440	42·03936 0	396 0	186·75604	3584	853·8332	2760	3976·923	1·940
2·65	10·03375	42·54906 2	397 6	189·79596	3612	871·2875	2792	4074·842	1·971
·66	10·11340	43·06274 0	399 0	192·87200	3640	889·0210	2824	4174·732	2·003
·67	10·19335	43·58040 8	400 4	195·98444	3668	907·0369	2858	4276·625	2·034
·68	10·27360	44·10208 0	402 0	199·13356	3695	925·3386	2891	4380·552	2·067
·69	10·35415	44·62777 2	403 6	202·31963	3724	943·9294	2925	4486·546	2·097
2·70	10·43500	45·15750 0	405 0	205·54294	3752	962·8127	2958	4594·637	2·131
·71	10·51615	45·69127 8	406 4	208·80377	3781	981·9918	2992	4704·859	2·165
·72	10·59760	46·22912 0	408 0	212·10241	3809	1001·4701	3027	4817·246	2·196
·73	10·67935	46·77104 2	409 6	215·43914	3838	1021·2511	3061	4931·829	2·232
·74	10·76140	47·31706 0	411 0	218·81425	3867	1041·3382	3097	5048·644	2·265
2·75	10·84375	47·86718 8	412 4	222·22803	3895	1061·7350	3130	5167·724	2·300
·76	10·92640	48·42144 0	414 0	225·68076	3924	1082·4448	3167	5289·104	2·334
·77	11·00935	48·97983 2	415 6	229·17273	3953	1103·4713	3203	5412·818	2·371
·78	11·09260	49·54238 0	417 0	232·70423	3983	1124·8181	3237	5538·903	2·405
·79	11·17615	50·10909 8	418 4	236·27556	4011	1146·4886	3274	5667·393	2·442
2·80	11·26000	50·68000 0	420 0	239·88700	4041	1168·4865	3310	5798·325	2·478
·81	11·34415	51·25510 2	421 6	243·53885	4071	1190·8154	3348	5931·735	2·516
·82	11·42860	51·83442 0	423 0	247·23141	4100	1213·4791	3384	6067·661	2·552
·83	11·51335	52·41796 8	424 4	250·96497	4130	1236·4812	3421	6206·139	2·591
·84	11·59840	53·00576 0	426 0	254·73983	4159	1259·8254	3459	6347·208	2·628
2·85	11·68375	53·59781 2	427 6	258·55628	4189	1283·5155	3496	6490·905	2·667
·86	11·76940	54·19414 0	429 0	262·41462	4220	1307·5552	3534	6637·269	2·706
·87	11·85535	54·79475 8	430 4	266·31516	4249	1331·9483	3573	6786·339	2·745
·88	11·94160	55·39968 0	432 0	270·25819	4279	1356·6987	3611	6938·154	2·785
·89	12·02815	56·00892 2	433 6	274·24401	4311	1381·8102	3650	7092·754	2·826
2·90	12·11500	56·62250 0	435 0	278·27294	4339	1407·2867	3690	7250·180	2·866
·91	12·20215	57·24042 8	436 4	282·34526	4372	1433·1322	3727	7410·472	2·907
·92	12·28960	57·86272 0	438 0	286·46130	4401	1459·3504	3769	7573·671	2·949
·93	12·37735	58·48939 2	439 6	290·62135	4432	1485·9455	3807	7739·819	2·991
·94	12·46540	59·12046 0	441 0	294·82572	4462	1512·9213	3849	7908·958	3·032
2·95	12·55375	59·75593 8	442 4	299·07471	4495	1540·2820	3888	8081·129	3·077
·96	12·64240	60·39584 0	444 0	303·36865	4525	1568·0315	3929	8256·377	3·119
·97	12·73135	61·04018 2	445 6	307·70784	4555	1596·1739	3971	8434·744	3·163
·98	12·82060	61·68898 0	447 0	312·09258	4588	1624·7134	4012	8616·274	3·207
2·99	12·91015	62·34224 8	448 4	316·52320	4618	1653·6541	4052	8801·011	3·252
3·00	13·00000	63·00000 0	450 0	321·00000	4650	1683·0000	4096	8989·000	3·297

† For $P_2(x)$, δ^2 is $+30$ throughout.

x	$P_2(x)$†	$P_3(x)$	δ^2	$P_4(x)$	δ^2	$P_5(x)$	δ^2	$P_6(x)$	δ^2
3·00	13·00000	63·00000	450	321·00000	4650	1683·0000	4096	8989·000	3·297
·01	13·09015	63·66225	452	325·52330	4682	1712·7555	4136	9180·286	3·343
·02	13·18060	64·32902	453	330·09342	4713	1742·9246	4180	9374·915	3·388
·03	13·27135	65·00032	454	334·71067	4745	1773·5117	4223	9572·932	3·436
·04	13·36240	65·67616	456	339·37537	4777	1804·5211	4265	9774·385	3·482
3·05	13·45375	66·35656	458	344·08784	4809	1835·9570	4308	9979·320	3·529
·06	13·54540	67·04154	459	348·84840	4840	1867·8237	4353	10187·784	3·578
·07	13·63735	67·73111	460	353·65736	4874	1900·1257	4396	10399·826	3·627
·08	13·72960	68·42528	462	358·51506	4905	1932·8673	4439	10615·495	3·674
·09	13·82215	69·12407	464	363·42181	4938	1966·0528	4486	10834·838	3·725
3·10	13·91500	69·82750	465	368·37794	4970	1999·6869	4529	11057·906	3·774
·11	14·00815	70·53558	466	373·38377	5003	2033·7739	4574	11284·748	3·824
·12	14·10160	71·24832	468	378·43963	5035	2068·3183	4620	11515·414	3·877
·13	14·19535	71·96574	470	383·54584	5069	2103·3247	4665	11749·957	3·926
·14	14·28940	72·68786	471	388·70274	5101	2138·7976	4711	11988·426	3·979
3·15	14·38375	73·41469	472	393·91065	5135	2174·7416	4758	12230·874	4·032
·16	14·47840	74·14624	474	399·16991	5167	2211·1614	4804	12477·354	4·083
·17	14·57335	74·88253	476	404·48084	5200	2248·0616	4851	12727·917	4·139
·18	14·66860	75·62358	477	409·84377	5235	2285·4469	4897	12982·619	4·191
·19	14·76415	76·36940	478	415·25905	5267	2323·3219	4946	13241·512	4·246
3·20	14·86000	77·12000	480	420·72700	5301	2361·6915	4993	13504·651	4·301
·21	14·95615	77·87540	482	426·24796	5335	2400·5604	5041	13772·091	4·357
·22	15·05260	78·63562	483	431·82227	5368	2439·9334	5089	14043·888	4·413
·23	15·14935	79·40067	484	437·45026	5403	2479·8153	5138	14320·098	4·468
·24	15·24640	80·17056	486	443·13228	5435	2520·2110	5187	14600·776	4·527
3·25	15·34375	80·94531	488	448·86865	5471	2561·1254	5236	14885·981	4·585
·26	15·44140	81·72494	489	454·65973	5505	2602·5634	5285	15175·771	4·641
·27	15·53935	82·50946	490	460·50586	5538	2644·5299	5336	15470·202	4·701
·28	15·63760	83·29888	492	466·40737	5574	2687·0300	5386	15769·334	4·761
·29	15·73615	84·09322	494	472·36462	5607	2730·0687	5435	16073·227	4·820
3·30	15·83500	84·89250	495	478·37794	5642	2773·6509	5488	16381·940	4·880
·31	15·93415	85·69673	496	484·44768	5678	2817·7819	5538	16695·533	4·943
·32	16·03360	86·50592	498	490·57420	5711	2862·4667	5588	17014·069	5·002
·33	16·13335	87·32009	500	496·75783	5746	2907·7103	5642	17337·607	5·065
·34	16·23340	88·13926	501	502·99892	5783	2953·5181	5693	17666·210	5·129
3·35	16·33375	88·96344	502	509·29784	5816	2999·8952	5746	17999·942	5·190
·36	16·43440	89·79264	504	515·65492	5853	3046·8469	5797	18338·864	5·256
·37	16·53535	90·62688	506	522·07053	5886	3094·3783	5852	18683·042	5·319
·38	16·63660	91·46618	507	528·54500	5924	3142·4949	5904	19032·539	5·385
·39	16·73815	92·31055	508	535·07871	5958	3191·2019	5957	19387·421	5·449
3·40	16·84000	93·16000	510	541·67200	5994	3240·5046	6013	19747·752	5·517
·41	16·94215	94·01455	512	548·32523	6030	3290·4086	6066	20113·600	5·583
·42	17·04460	94·87422	513	555·03876	6065	3340·9192	6121	20485·031	5·651
·43	17·14735	95·73902	514	561·81294	6102	3392·0419	6175	20862·113	5·717
·44	17·25040	96·60896	516	568·64814	6137	3443·7821	6231	21244·912	5·788
3·45	17·35375	97·48406	518	575·54471	6175	3496·1454	6287	21633·499	5·856
·46	17·45740	98·36434	519	582·50303	6209	3549·1374	6342	22027·942	5·926
·47	17·56135	99·24981	520	589·52344	6247	3602·7636	6398	22428·311	5·997
·48	17·66560	100·14048	522	596·60632	6283	3657·0296	6456	22834·677	6·067
·49	17·77015	101·03637	524	603·75203	6320	3711·9412	6511	23247·110	6·140
3·50	17·87500	101·93750	525	610·96094	6356	3767·5039	6570	23665·683	6·211

† For $P_2(x)$, δ^2 is $+30$ throughout.

LEGENDRE POLYNOMIALS

x	$P_2(x)$†	$P_3(x)$	δ^2	$P_4(x)$	δ^2	$P_5(x)$	δ^2	$P_6(x)$	δ^2
3·50	17·87500	101·93750	525	610·9609	636	3767·504	657	23665·683	6·211
·51	17·98015	102·84388	526	618·2334	639	3823·724	662	24090·467	6·284
·52	18·08560	103·75552	528	625·5698	643	3880·606	669	24521·535	6·358
·53	18·19135	104·67244	530	632·9705	647	3938·157	674	24958·961	6·433
·54	18·29740	105·59466	531	640·4359	650	3996·382	680	25402·820	6·507
3·55	18·40375	106·52219	532	647·9663	654	4055·287	686	25853·186	6·583
·56	18·51040	107·45504	534	655·5621	658	4114·878	692	26310·135	6·658
·57	18·61735	108·39323	536	663·2237	662	4175·161	698	26773·742	6·736
·58	18·72460	109·33678	537	670·9515	665	4236·142	704	27244·085	6·813
·59	18·83215	110·28570	538	678·7458	669	4297·827	709	27721·241	6·891
3·60	18·94000	111·24000	540	686·6070	673	4360·221	717	28205·288	6·971
·61	19·04815	112·19970	542	694·5355	677	4423·332	722	28696·306	7·049
·62	19·15660	113·16482	543	702·5317	680	4487·165	727	29194·373	7·129
·63	19·26535	114·13537	544	710·5959	685	4551·725	735	29699·569	7·211
·64	19·37440	115·11136	546	718·7286	687	4617·020	741	30211·976	7·292
3·65	19·48375	116·09281	548	726·9300	693	4683·056	746	30731·675	7·374
·66	19·59340	117·07974	549	735·2007	695	4749·838	754	31258·748	7·457
·67	19·70335	118·07216	550	743·5409	700	4817·374	758	31793·278	7·540
·68	19·81360	119·07008	552	751·9511	704	4885·668	766	32335·348	7·626
·69	19·92415	120·07352	554	760·4317	706	4954·728	772	32885·044	7·709
3·70	20·03500	121·08250	555	768·9829	712	5024·560	779	33442·449	7·795
·71	20·14615	122·09703	556	777·6053	716	5095·171	784	34007·649	7·882
·72	20·25760	123·11712	558	786·2993	718	5166·566	792	34580·731	7·969
·73	20·36935	124·14279	560	795·0651	723	5238·753	797	35161·782	8·057
·74	20·48140	125·17406	561	803·9032	727	5311·737	805	35750·890	8·145
3·75	20·59375	126·21094	562	812·8140	730	5385·526	810	36348·143	8·234
·76	20·70640	127·25344	564	821·7978	736	5460·125	818	36953·630	8·325
·77	20·81935	128·30158	566	830·8552	738	5535·542	824	37567·442	8·414
·78	20·93260	129·35538	567	839·9864	743	5611·783	831	38189·668	8·507
·79	21·04615	130·41485	568	849·1919	746	5688·855	837	38820·401	8·599
3·80	21·16000	131·48000	570	858·4720	751	5766·764	845	39459·733	8·691
·81	21·27415	132·55085	572	867·8272	755	5845·518	851	40107·756	8·785
·82	21·38860	133·62742	573	877·2579	758	5925·123	858	40764·564	8·879
·83	21·50335	134·70972	574	886·7644	762	6005·586	864	41430·251	8·975
·84	21·61840	135·79776	576	896·3471	768	6086·913	872	42104·913	9·071
3·85	21·73375	136·89156	578	906·0066	770	6169·112	879	42788·646	9·167
·86	21·84940	137·99114	579	915·7431	775	6252·190	885	43481·546	9·265
·87	21·96535	139·09651	580	925·5571	778	6336·153	893	44183·711	9·363
·88	22·08160	140·20768	582	935·4489	784	6421·009	900	44895·239	9·461
·89	22·19815	141·32467	584	945·4191	786	6506·765	906	45616·228	9·562
3·90	22·31500	142·44750	585	955·4679	792	6593·427	914	46346·779	9·662
·91	22·43215	143·57618	586	965·5959	794	6681·003	921	47086·992	9·764
·92	22·54960	144·71072	588	975·8033	800	6769·500	928	47836·969	9·865
·93	22·66735	145·85114	590	986·0907	803	6858·925	935	48596·811	9·969
·94	22·78540	146·99746	591	996·4584	808	6949·285	943	49366·622	10·072
3·95	22·90375	148·14969	592	1006·9069	811	7040·588	950	50146·505	10·176
·96	23·02240	149·30784	594	1017·4365	816	7132·841	958	50936·564	10·283
·97	23·14135	150·47193	596	1028·0477	820	7226·052	964	51736·906	10·387
·98	23·26060	151·64198	597	1038·7409	825	7320·227	972	52547·635	10·496
3·99	23·38015	152·81800	598	1049·5166	827	7415·374	979	53368·860	10·603
4·00	23·50000	154·00000	600	1060·3750	833	7511·500	988	54200·688	10·710

† For $P_2(x)$, δ^2 is $+30$ throughout.

x	$P_2(x)$†	$P_3(x)$	δ^2	$P_4(x)$	δ^2	$P_5(x)$	δ^2	$P_6(x)$	δ^2
4·00	23·50000	154·00000	600	1060·3750	833	7511·500	988	54200·69	10·71
·01	23·62015	155·18800	602	1071·3167	837	7608·614	994	55043·23	10·82
·02	23·74060	156·38202	603	1082·3421	840	7706·722	1·002	55896·59	10·93
·03	23·86135	157·58207	604	1093·4515	846	7805·832	1·009	56760·88	11·04
·04	23·98240	158·78816	606	1104·6455	849	7905·951	1·019	57636·21	11·16
4·05	24·10375	160·00031	608	1115·9244	854	8007·089	1·024	58522·70	11·26
·06	24·22540	161·21854	609	1127·2887	857	8109·251	1·033	59420·45	11·38
·07	24·34735	162·44286	610	1138·7387	863	8212·446	1·040	60329·58	11·50
·08	24·46960	163·67328	612	1150·2750	866	8316·681	1·049	61250·21	11·61
·09	24·59215	164·90982	614	1161·8979	871	8421·965	1·056	62182·45	11·73
4·10	24·71500	166·15250	615	1173·6079	875	8528·305	1·063	63126·42	11·83
·11	24·83815	167·40133	616	1185·4054	880	8635·708	1·073	64082·22	11·97
·12	24·96160	168·65632	618	1197·2909	883	8744·184·	1·080	65049·99	12·08
·13	25·08535	169·91749	620	1209·2647	888	8853·740	1·087	66029·84	12·20
·14	25·20940	171·18486	621	1221·3273	892	8964·383	1·096	67021·89	12·32
4·15	25·33375	172·45844	622	1233·4791	897	9076·122	1·104	68026·26	12·44
·16	25·45840	173·73824	624	1245·7206	901	9188·965	1·112	69043·07	12·57
·17	25·58335	175·02428	626	1258·0522	905	9302·920	1·121	70072·45	12·68
·18	25·70860	176·31658	627	1270·4743	911	9417·996	1·127	71114·51	12·81
·19	25·83415	177·61515	628	1282·9875	913	9534·199	1·138	72169·38	12·94
4·20	25·96000	178·92000	630	1295·5920	919	9651·540	1·144	73237·19	13·07
·21	26·08615	180·23115	632	1308·2884	923	9770·025	1·153	74318·07	13·18
·22	26·21260	181·54862	633	1321·0771	928	9889·663	1·162	75412·13	13·31
·23	26·33935	182·87242	634	1333·9586	931	10010·463	1·169	76519·50	13·46
·24	26·46640	184·20256	636	1346·9332	937	10132·432	1·179	77640·33	13·57
4·25	26·59375	185·53906	638	1360·0015	940	10255·580	1·187	78774·73	13·70
·26	26·72140	186·88194	639	1373·1638	946	10379·915	1·194	79922·83	13·84
·27	26·84935	188·23121	640	1386·4207	949	10505·444	1·205	81084·77	13·97
·28	26·97760	189·58688	642	1399·7725	955	10632·178	1·212	82260·68	14·10
·29	27·10615	190·94897	644	1413·2198	958	10760·124	1·221	83450·69	14·24
4·30	27·23500	192·31750	645	1426·7629	964	10889·291	1·230	84654·94	14·37
·31	27·36415	193·69248	646	1440·4024	967	11019·688	1·238	85873·56	14·52
·32	27·49360	195·07392	648	1454·1386	973	11151·323	1·247	87106·70	14·63
·33	27·62335	196·46184	650	1467·9721	976	11284·205	1·256	88354·47	14·79
·34	27·75340	197·85626	651	1481·9032	982	11418·343	1·265	89617·03	14·92
4·35	27·88375	199·25719	652	1495·9325	986	11553·746	1·273	90894·51	15·07
·36	28·01440	200·66464	654	1510·0604	990	11690·422	1·283	92187·06	15·20
·37	28·14535	202·07863	656	1524·2873	996	11828·381	1·292	93494·81	15·34
·38	28·27660	203·49918	657	1538·6138	999	11967·632	1·299	94817·90	15·50
·39	28·40815	204·92630	658	1553·0402	1004	12108·182	1·311	96156·49	15·62
4·40	28·54000	206·36000	660	1567·5670	1009	12250·043	1·317	97510·70	15·79
·41	28·67215	207·80030	662	1582·1947	1014	12393·221	1·329	98880·70	15·92
·42	28·80460	209·24722	663	1596·9238	1018	12537·728	1·336	1 00266·62	16·07
·43	28·93735	210·70077	664	1611·7547	1023	12683·571	1·347	1 01668·61	16·22
·44	29·07040	212·16096	666	1626·6879	1027	12830·761	1·354	1 03086·82	16·37
4·45	29·20375	213·62781	668	1641·7238	1032	12979·305	1·365	1 04521·40	16·51
·46	29·33740	215·10134	669	1656·8629	1037	13129·214	1·374	1 05972·49	16·67
·47	29·47135	216·58156	670	1672·1057	1042	13280·497	1·384	1 07440·25	16·83
·48	29·60560	218·06848	672	1687·4527	1046	13433·164	1·392	1 08924·84	16·97
·49	29·74015	219·56212	674	1702·9043	1050	13587·223	1·402	1 10426·40	17·13
4·50	29·87500	221·06250	675	1718·4609	1057	13742·684	1·411	1 11945·09	17·28

† For $P_2(x)$, δ^2 is $+30$ throughout.

x	$P_2(x)$†	$P_3(x)$	δ^2	$P_4(x)$	δ^2	$P_5(x)$	δ^2	$P_6(x)$	δ^2
4·50	29·87500	221·06250	675	1718·4609	1057	13742·684	1·411	1 11945·09	17·28
·51	30·01015	222·56963	676	1734·1232	1059	13899·556	1·422	1 13481·06	17·44
·52	30·14560	224·08352	678	1749·8914	1066	14057·850	1·430	1 15034·47	17·60
·53	30·28135	225·60419	680	1765·7662	1070	14217·574	1·441	1 16605·48	17·76
·54	30·41740	227·13166	681	1781·7480	1074	14378·739	1·450	1 18194·25	17·91
4·55	30·55375	228·66594	682	1797·8372	1080	14541·354	1·460	1 19800·93	18·08
·56	30·69040	230·20704	684	1814·0344	1084	14705·429	1·469	1 21425·69	18·23
·57	30·82735	231·75498	686	1830·3400	1088	14870·973	1·479	1 23068·68	18·41
·58	30·96460	233·30978	687	1846·7544	1095	15037·996	1·489	1 24730·08	18·55
·59	31·10215	234·87145	688	1863·2783	1098	15206·508	1·499	1 26410·03	18·74
4·60	31·24000	236·44000	690	1879·9120	1104	15376·519	1·510	1 28108·72	18·89
·61	31·37815	238·01545	692	1896·6561	1107	15548·040	1·518	1 29826·30	19·06
·62	31·51660	239·59782	693	1913·5109	1114	15721·079	1·529	1 31562·94	19·24
·63	31·65535	241·18712	694	1930·4771	1118	15895·647	1·538	1 33318·82	19·39
·64	31·79440	242·78336	696	1947·5551	1122	16071·753	1·550	1 35094·09	19·57
4·65	31·93375	244·38656	698	1964·7453	1129	16249·409	1·559	1 36888·93	19·74
·66	32·07340	245·99674	699	1982·0484	1131	16428·624	1·570	1 38703·51	19·91
·67	32·21335	247·61391	700	1999·4646	1139	16609·409	1·579	1 40538·00	20·09
·68	32·35360	249·23808	702	2016·9947	1141	16791·773	1·590	1 42392·58	20·26
·69	32·49415	250·86927	704	2034·6389	1148	16975·727	1·600	1 44267·42	20·44
4·70	32·63500	252·50750	705	2052·3979	1153	17161·281	1·610	1 46162·70	20·62
·71	32·77615	254·15278	706	2070·2722	1156	17348·445	1·622	1 48078·60	20·78
·72	32·91760	255·80512	708	2088·2621	1162	17537·231	1·631	1 50015·28	20·98
·73	33·05935	257·46454	710	2106·3682	1168	17727·648	1·641	1 51972·94	21·16
·74	33·20140	259·13106	711	2124·5911	1172	17919·706	1·654	1 53951·76	21·32
4·75	33·34375	260·80469	712	2142·9312	1176	18113·418	1·662	1 55951·90	21·53
·76	33·48640	262·48544	714	2161·3889	1183	18308·792	1·674	1 57973·57	21·69
·77	33·62935	264·17333	716	2179·9649	1186	18505·840	1·684	1 60016·93	21·89
·78	33·77260	265·86838	717	2198·6595	1193	18704·572	1·695	1 62082·18	22·08
·79	33·91615	267·57060	718	2217·4734	1197	18904·999	1·706	1 64169·51	22·25
4·80	34·06000	269·28000	720	2236·4070	1202	19107·132	1·717	1 66279·09	22·46
·81	34·20415	270·99660	722	2255·4608	1207	19310·982	1·727	1 68411·13	22·63
·82	34·34860	272·72042	723	2274·6353	1212	19516·559	1·739	1 70565·80	22·83
·83	34·49335	274·45147	724	2293·9310	1218	19723·875	1·749	1 72743·30	23·03
·84	34·63840	276·18976	726	2313·3485	1222	19932·940	1·760	1 74943·83	23·21
4·85	34·78375	277·93531	728	2332·8882	1227	20143·765	1·772	1 77167·57	23·42
·86	34·92940	279·68814	729	2352·5506	1233	20356·362	1·782	1 79414·73	23·60
·87	35·07535	281·44826	730	2372·3363	1237	20570·741	1·794	1 81685·49	23·80
·88	35·22160	283·21568	732	2392·2457	1243	20786·914	1·804	1 83980·05	24·01
·89	35·36815	284·99042	734	2412·2794	1248	21004·891	1·817	1 86298·62	24·20
4·90	35·51500	286·77250	735	2432·4379	1253	21224·685	1·826	1 88641·39	24·40
·91	35·66215	288·56193	736	2452·7217	1259	21446·305	1·840	1 91008·56	24·60
·92	35·80960	290·35872	738	2473·1314	1262	21669·765	1·849	1 93400·33	24·82
·93	35·95735	292·16289	740	2493·6673	1270	21895·074	1·861	1 95816·92	25·01
·94	36·10540	293·97446	741	2514·3302	1272	22122·244	1·873	1 98258·52	25·21
4·95	36·25375	295·79344	742	2535·1203	1280	22351·287	1·885	2 00725·33	25·44
·96	36·40240	297·61984	744	2556·0384	1284	22582·215	1·895	2 03217·58	25·62
·97	36·55135	299·45368	746	2577·0849	1289	22815·038	1·908	2 05735·45	25·86
·98	36·70060	301·29498	747	2598·2603	1295	23049·769	1·919	2 08279·18	26·04
4·99	36·85015	303·14375	748	2619·5652	1299	23286·419	1·931	2 10848·95	26·28
5·00	37·00000	305·00000	750	2641·0000	1305	23525·000	1·942	2 13445·00	26·48

† For $P_2(x)$, δ^2 is $+30$ throughout.

x	$P_2(x)$†	$P_3(x)$	δ^2	$P_4(x)$	δ^2	$P_5(x)$	δ^2	$P_6(x)$	δ^2
5·00	37·00000	305·00000	750	2641·0000	1305	23525·000	1·942	2 13445·00	26·48
·01	37·15015	306·86375	752	2662·5653	1311	23765·523	1·955	2 16067·53	26·69
·02	37·30060	308·73502	753	2684·2617	1315	24008·001	1·965	2 18716·75	26·91
·03	37·45135	310·61382	754	2706·0896	1321	24252·444	1·979	2 21392·88	27·14
·04	37·60240	312·50016	756	2728·0496	1326	24498·866	1·989	2 24096·15	27·34
5·05	37·75375	314·39406	758	2750·1422	1332	24747·277	2·003	2 26826·76	27·57
·06	37·90540	316·29554	759	2772·3680	1336	24997·691	2·013	2 29584·94	27·79
·07	38·05735	318·20461	760	2794·7274	1342	25250·118	2·027	2 32370·91	28·01
·08	38·20960	320·12128	762	2817·2210	1347	25504·572	2·037	2 35184·89	28·24
·09	38·36215	322·04557	764	2839·8493	1353	25761·063	2·051	2 38027·11	28·46
5·10	38·51500	323·97750	765	2862·6129	1359	26019·605	2·062	2 40897·79	28·69
·11	38·66815	325·91708	766	2885·5124	1362	26280·209	2·075	2 43797·16	28·92
·12	38·82160	327·86432	768	2908·5481	1369	26542·888	2·087	2 46725·45	29·14
·13	38·97535	329·81924	770	2931·7207	1375	26807·654	2·099	2 49682·88	29·38
·14	39·12940	331·78186	771	2955·0308	1379	27074·519	2·113	2 52669·69	29·61
5·15	39·28375	333·75219	772	2978·4788	1385	27343·497	2·123	2 55686·11	29·85
·16	39·43840	335·73024	774	3002·0653	1390	27614·598	2·137	2 58732·38	30·07
·17	39·59335	337·71603	776	3025·7908	1396	27887·836	2·150	2 61808·72	30·31
·18	39·74860	339·70958	777	3049·6559	1401	28163·224	2·161	2 64915·37	30·55
·19	39·90415	341·71090	778	3073·6611	1407	28440·773	2·176	2 68052·57	30·80
5·20	40·06000	343·72000	780	3097·8070	1412	28720·498	2·186	2 71220·57	31·03
·21	40·21615	345·73690	782	3122·0941	1417	29002·409	2·200	2 74419·60	31·27
·22	40·37260	347·76162	783	3146·5229	1424	29286·520	2·214	2 77649·90	31·51
·23	40·52935	349·79417	784	3171·0941	1428	29572·845	2·224	2 80911·71	31·77
·24	40·68640	351·83456	786	3195·8081	1434	29861·394	2·240	2 84205·29	32·00
5·25	40·84375	353·88281	788	3220·6655	1440	30152·183	2·251	2 87530·87	32·26
·26	41·00140	355·93894	789	3245·6669	1445	30445·223	2·265	2 90888·71	32·50
·27	41·15935	358·00296	790	3270·8128	1450	30740·528	2·277	2 94279·05	32·76
·28	41·31760	360·07488	792	3296·1037	1456	31038·110	2·290	2 97702·15	33·00
·29	41·47615	362·15472	794	3321·5402	1462	31337·982	2·305	3 01158·25	33·26
5·30	41·63500	364·24250	795	3347·1229	1468	31640·159	2·316	3 04647·61	33·51
·31	41·79415	366·33823	796	3372·8524	1472	31944·652	2·331	3 08170·48	33·77
·32	41·95360	368·44192	798	3398·7291	1478	32251·476	2·343	3 11727·12	34·03
·33	42·11335	370·55359	800	3424·7536	1485	32560·643	2·358	3 15317·79	34·29
·34	42·27340	372·67326	801	3450·9266	1489	32872·168	2·369	3 18942·75	34·54
5·35	42·43375	374·80094	802	3477·2485	1495	33186·062	2·384	3 22602·25	34·81
·36	42·59440	376·93664	804	3503·7199	1501	33502·340	2·397	3 26296·56	35·08
·37	42·75535	379·08038	806	3530·3414	1506	33821·015	2·412	3 30025·95	35·33
·38	42·91660	381·23218	807	3557·1135	1513	34142·102	2·423	3 33790·67	35·60
·39	43·07815	383·39205	808	3584·0369	1517	34465·612	2·439	3 37590·99	35·88
5·40	43·24000	385·56000	810	3611·1120	1524	34791·561	2·451	3 41427·19	36·14
·41	43·40215	387·73605	812	3638·3395	1528	35119·961	2·466	3 45299·53	36·41
·42	43·56460	389·92022	813	3665·7198	1536	35450·827	2·479	3 49208·28	36·69
·43	43·72735	392·11252	814	3693·2537	1540	35784·172	2·493	3 53153·72	36·95
·44	43·89040	394·31296	816	3720·9416	1546	36120·010	2·507	3 57136·11	37·24
5·45	44·05375	396·52156	818	3748·7841	1552	36458·355	2·521	3 61155·74	37·51
·46	44·21740	398·73834	819	3776·7818	1557	36799·221	2·535	3 65212·88	37·79
·47	44·38135	400·96331	820	3804·9352	1564	37142·622	2·549	3 69307·81	38·08
·48	44·54560	403·19648	822	3833·2450	1569	37488·572	2·563	3 73440·82	38·34
·49	44·71015	405·43787	824	3861·7117	1575	37837·085	2·578	3 77612·17	38·64
5·50	44·87500	407·68750	825	3890·3359	1581	38188·176	2·591	3 81822·16	38·92

† For $P_2(x)$, δ^2 is $+30$ throughout.

x	$P_2(x)$†	$P_3(x)$	δ^2	$P_4(x)$	δ^2	$P_5(x)$	δ^2	$P_6(x)$	δ^2
5·50	44·87500	407·68750	825	3890·3359	1581	38188·176	2·591	3 81822·16	38·92
·51	45·04015	409·94538	826	3919·1182	1586	38541·858	2·606	3 86071·07	39·21
·52	45·20560	412·21152	828	3948·0591	1592	38898·146	2·620	3 90359·19	39·49
·53	45·37135	414·48594	830	3977·1592	1598	39257·054	2·634	3 94686·80	39·79
·54	45·53740	416·76866	831	4006·4191	1604	39618·596	2·650	3 99054·20	40·07
5·55	45·70375	419·05969	832	4035·8394	1610	39982·788	2·663	4 03461·67	40·37
·56	45·87040	421·35904	834	4065·4207	1615	40349·643	2·677	4 07909·51	40·66
·57	46·03735	423·66673	836	4095·1635	1621	40719·175	2·694	4 12398·01	40·96
·58	46·20460	425·98278	837	4125·0684	1627	41091·401	2·706	4 16927·47	41·26
·59	46·37215	428·30720	838	4155·1360	1634	41466·333	2·722	4 21498·19	41·55
5·60	46·54000	430·64000	840	4185·3670	1638	41843·987	2·737	4 26110·46	41·86
·61	46·70815	432·98120	842	4215·7618	1646	42224·378	2·751	4 30764·59	42·16
·62	46·87660	435·33082	843	4246·3212	1650	42607·520	2·767	4 35460·88	42·46
·63	47·04535	437·68887	844	4277·0456	1656	42993·429	2·780	4 40199·63	42·77
·64	47·21440	440·05536	846	4307·9356	1663	43382·118	2·796	4 44981·15	43·08
5·65	47·38375	442·43031	848	4338·9919	1668	43773·603	2·812	4 49805·75	43·38
·66	47·55340	444·81374	849	4370·2150	1675	44167·900	2·825	4 54673·73	43·69
·67	47·72335	447·20566	850	4401·6056	1680	44565·022	2·842	4 59585·40	44·02
·68	47·89360	449·60608	852	4433·1642	1687	44964·986	2·856	4 64541·09	44·31
·69	48·06415	452·01502	854	4464·8915	1691	45367·806	2·872	4 69541·09	44·64
5·70	48·23500	454·43250	855	4496·7879	1699	45773·498	2·887	4 74585·73	44·96
·71	48·40615	456·85853	856	4528·8542	1704	46182·077	2·902	4 79675·33	45·27
·72	48·57760	459·29312	858	4561·0909	1711	46593·558	2·917	4 84810·20	45·59
·73	48·74935	461·73629	860	4593·4987	1715	47007·956	2·934	4 89990·66	45·92
·74	48·92140	464·18806	861	4626·0780	1723	47425·288	2·948	4 95217·04	46·25
5·75	49·09375	466·64844	862	4658·8296	1728	47845·568	2·963	5 00489·67	46·55
·76	49·26640	469·11744	864	4691·7540	1734	48268·811	2·981	5 05808·85	46·91
·77	49·43935	471·59508	866	4724·8518	1741	48695·035	2·995	5 11174·94	47·21
·78	49·61260	474·08138	867	4758·1237	1746	49124·254	3·011	5 16588·24	47·56
·79	49·78615	476·57635	868	4791·5702	1753	49556·484	3·026	5 22049·10	47·89
5·80	49·96000	479·08000	870	4825·1920	1758	49991·740	3·044	5 27557·85	48·21
·81	50·13415	481·59235	872	4858·9896	1765	50430·040	3·057	5 33114·81	48·57
·82	50·30860	484·11342	873	4892·9637	1771	50871·397	3·075	5 38720·34	48·89
·83	50·48335	486·64322	874	4927·1149	1777	51315·829	3·091	5 44374·76	49·24
·84	50·65840	489·18176	876	4961·4438	1783	51763·352	3·105	5 50078·42	49·57
5·85	50·83375	491·72906	878	4995·9510	1789	52213·980	3·124	5 55831·65	49·92
·86	51·00940	494·28514	879	5030·6371	1795	52667·732	3·137	5 61634·80	50·27
·87	51·18535	496·85001	880	5065·5027	1802	53124·621	3·156	5 67488·22	50·60
·88	51·36160	499·42368	882	5100·5485	1807	53584·666	3·171	5 73392·24	50·97
·89	51·53815	502·00617	884	5135·7750	1814	54047·882	3·187	5 79347·23	51·31
5·90	51·71500	504·59750	885	5171·1829	1821	54514·285	3·204	5 85353·53	51·66
·91	51·89215	507·19768	886	5206·7729	1825	54983·892	3·220	5 91411·49	52·01
·92	52·06960	509·80672	888	5242·5454	1833	55456·719	3·236	5 97521·46	52·38
·93	52·24735	512·42464	890	5278·5012	1839	55932·782	3·254	6 03683·81	52·73
·94	52·42540	515·05146	891	5314·6409	1844	56412·099	3·270	6 09898·89	53·09
5·95	52·60375	517·68719	892	5350·9650	1852	56894·686	3·286	6 16167·06	53·45
·96	52·78240	520·33184	894	5387·4743	1857	57380·559	3·303	6 22488·68	53·81
·97	52·96135	522·98543	896	5424·1693	1864	57869·735	3·320	6 28864·11	54·18
·98	53·14060	525·64798	897	5461·0507	1869	58362·231	3·337	6 35293·72	54·54
5·99	53·32015	528·31950	898	5498·1190	1877	58858·064	3·353	6 41777·87	54·92
6·00	53·50000	531·00000	900	5535·3750	1882	59357·250	3·371	6 48316·94	55·28

† For $P_2(x)$, δ^2 is $+30$ throughout.

x	$P_7(x)$	$*\delta^2$	$P_8(x)$	$*\delta^2$	$P_9(x)$	$*\delta^2$
0·00	− 0·00000 00	0	+ 0·27343 75	− 197 00	+ 0·00000 00	0
·01	·02185 53	+ 11 81	·27245 37	196 36	·02457 33	− 21 64
·02	·04359 26	23 55	·26950 87	194 42	·04893 05	43 14
·03	·06509 45	35 23	·26462 19	191 17	·07285 70	64 25
·04	·08624 44	46 70	·25782 57	186 68	·09614 19	84 91
0·05	− 0·10692 76	+ 58 02	+ 0·24916 50	− 180 92	+ 0·11857 90	− 104 87
·06	·12703 11	69 03	·23869 73	173 98	·13996 89	124 02
·07	·14644 48	79 78	·22649 20	165 81	·16012 04	142 18
·08	·16506 14	90 14	·21263 06	156 57	·17885 21	159 24
·09	·18277 73	100 12	·19720 56	146 18	·19599 37	175 03
0·10	− 0·19949 29	+ 109 60	+ 0·18032 07	− 134 79	+ 0·21138 76	− 189 38
·11	·21511 33	118 67	·16208 97	122 44	·22489 04	202 27
·12	·22954 81	127 10	·14263 60	109 13	·23637 36	213 46
·13	·24271 29	135 02	·12209 24	95 08	·24572 53	222 97
·14	·25452 87	142 29	·10059 95	80 20	·25285 07	230 59
0·15	− 0·26492 29	+ 148 86	+ 0·07830 58	− 64 69	+ 0·25767 37	− 236 33
·16	·27382 97	154 79	·05536 63	48 59	·26013 71	240 07
·17	·28119 00	159 92	·03194 18	31 98	·26020 36	241 76
·18	·28695 25	164 31	+ ·00819 82	− 15 01	·25785 63	241 36
·19	·29107 34	167 91	− ·01569 49	+ 2 29	·25309 92	238 90
0·20	− 0·29351 68	+ 170 64	− 0·03956 48	+ 19 74	+ 0·24595 71	− 234 26
·21	·29425 54	172 52	·06323 71	37 29	·23647 63	227 52
·22	·29327 04	173 54	·08653 65	54 84	·22472 41	218 70
·23	·29055 17	173 63	·10928 78	72 23	·21078 87	207 79
·24	·28609 84	172 81	·13131 73	89 34	·19477 91	194 87
0·25	− 0·27991 87	+ 171 05	− 0·15245 40	+ 106 13	+ 0·17682 44	− 180 00
·26	·27203 02	168 34	·17253 03	122 42	·15707 31	163 30
·27	·26246 00	164 70	·19138 35	138 10	·13569 21	144 77
·28	·25124 46	160 05	·20885 70	153 10	·11286 64	124 64
·29	·23843 04	154 50	·22480 11	167 23	·08879 71	102 99
0·30	− 0·22407 30	+ 147 94	− 0·23907 46	+ 180 44	+ 0·06370 04	− 79 96
·31	·20823 79	140 47	·25154 57	192 57	·03780 63	55 71
·32	·19099 99	132 02	·26209 32	203 56	+ ·01135 69	30 48
·33	·17244 34	122 66	·27060 75	213 25	− ·01539 57	− 4 36
·34	·15266 19	112 43	·27699 18	221 60	·04219 08	+ 22 33
0·35	− 0·13175 78	+ 101 25	− 0·28116 29	+ 228 43	− 0·06876 18	+ 49 47
·36	·10984 27	89 26	·28305 26	233 71	·09483 78	76 78
·37	·08703 65	76 47	·28260 83	237 30	·12014 61	104 03
·38	·06346 71	62 83	·27979 41	239 21	·14441 47	130 94
·39	·03927 07	48 52	·27459 12	239 24	·16737 49	157 30
0·40	− 0·01459 04	+ 33 47	− 0·26699 93	+ 237 42	− 0·18876 36	+ 182 83
·41	+ ·01042 35	17 82	·25703 68	233 64	·20832 61	207 21
·42	·03561 45	+ 1 55	·24474 16	227 86	·22581 90	230 23
·43	·06082 01	− 15 18	·23017 15	220 08	·24101 27	251 57
·44	·08587 30	32 42	·21340 44	210 21	·25369 43	270 98
0·45	+ 0·11060 11	− 49 96	− 0·19453 90	+ 198 29	− 0·26367 02	+ 288 13
·46	·13482 90	67 85	·17369 46	184 25	·27076 93	302 83
·47	·15837 81	85 90	·15101 15	168 18	·27484 52	314 75
·48	·18106 80	104 06	·12665 04	150 05	·27577 91	323 67
·49	·20271 73	122 23	·10079 25	129 93	·27348 23	329 33
0·50	+ 0·22314 45	− 140 29	− 0·07363 89	+ 107 85	− 0·26789 86	+ 331 50

* These differences have been modified; see page A 4.

x	$P_7(x)$	$*\delta^2$	δ^4	$P_8(x)$	$*\delta^2$	δ^4	$P_9(x)$	$*\delta^2$	δ^4
0·50	+0·22314 45	− 140 29		− 0·07363 89	+ 107 85		− 0·26789 86	+ 331 50	
·51	·24216 92	158 16		·04541 02	83 92		·25900 67	329 97	
·52	·25961 30	175 65		− ·01634 56	58 21		·24682 22	324 58	
·53	·27530 11	192 70		+ ·01329 80	30 81		·23139 94	315 11	
·54	·28906 33	209 16		·04324 70	+ 1 95		·21283 32	301 47	
0·55	+0·30073 53	− 224 86		+0·07321 29	− 28 36		− 0·19126 02	+ 283 52	
·56	·31016 03	239 68		·10289 31	59 79		·16686 00	261 24	
·57	·31719 04	253 48		·13197 35	92 26		·13985 55	234 53	
·58	·32168 80	266 02		·16012 99	125 47		·11051 37	203 48	
·59	·32352 79	277 23		·18703 07	159 19		·07914 50	168 10	
0·60	+0·32259 84	− 286 85		+0·21233 92	− 193 15		− 0·04610 30	+ 128 50	
·61	·31880 36	294 72		·23571 64	227 01		− ·01178 33	84 91	
·62	·31206 51	300 69		·25682 43	260 44		+ ·02337 86	+ 37 54	
·63	·30232 37	304 47		·27532 92	293 11		·05890 95	− 13 33	
·64	·28954 19	305 90		·29090 52	324 56		·09430 14	67 30	
0·65	+0·27370 58	− 304 79		+0·30323 86	− 354 40		+0·12901 55	− 123 89	
·66	·25482 70	300 83		·31203 19	382 09		·16248 69	182 59	
·67	·23294 55	293 85		·31700 90	407 25		·19412 98	242 73	
·68	·20813 16	283 56		·31791 95	429 18		·22334 41	303 60	
·69	·18048 86	269 73		·31454 50	447 43		·24952 27	364 34	
0·70	+0·15015 53	− 252 11		+0·30670 43	− 460 36	+ 5 01	+0·27205 99	− 424 00	
·71	·11730 85	230 36		·29426 00	469 09	5 77	·29036 11	481 53	
·72	·08216 61	204 27		·27712 48	472 05	6 46	·30385 32	535 69	
·73	·04498 96	173 52		·25526 91	468 55	7 31	·31199 70	585 22	
·74	+ ·00608 71	137 78		·22872 79	457 74	8 13	·31430 00	628 53	
0·75	− 0·03418 35	− 96 77		+0·19760 93	− 438 80	+ 9 01	+0·31033 19	− 662 40	+ 9 63
·76	·07541 15	− 50 15		·16210 27	410 85	10 01	·29973 98	688 06	11 50
·77	·11713 01	+ 2 38		·12248 76	372 89	10 92	·28226 71	702 22	13 66
·78	·15881 33	61 22		·07914 36	324 01	12 06	·25777 22	702 72	15 95
·79	·19987 21	126 68		+ ·03255 95	263 07	13 10	·22625 01	687 27	18 15
0·80	− 0·23965 12	+ 200 50	+ 7 38	− 0·01665 53	− 189 03	+14 28	+0·18785 53	− 653 37	+ 21 17
·81	·27742 53	280 41	7 75	·06776 04	− 100 71	15 50	·14292 68	598 30	24 08
·82	·31239 53	368 07	8 18	·11987 26	+ 3 11	16 81	·09201 53	519 15	27 20
·83	·34368 46	463 91	8 53	·17195 37	123 74	18 07	+ ·03591 23	412 80	30 60
·84	·37033 48	568 28	9 01	·22279 74	262 44	19 55	− ·02431 87	275 85	34 21
0·85	− 0·39130 22	+ 681 66	+ 9 43	− 0·27101 67	+ 420 69	+21 00	− 0·08730 82	− 104 69	+* 38 09
·86	·40545 30	804 47	9 82	·31502 91	599 94	22 51	·15134 46	+ 104 56	42 18
·87	·41155 91	937 10	10 34	·35304 21	801 70	24 11	·21433 54	355 99	46 69
·88	·40829 42	1080 07	10 76	·38303 81	1027 57	25 84	·27376 63	654 11	51 31
·89	·39422 86	1233 80	11 27	·40275 84	1279 28	27 50	·32665 61	1003 54	56 37
0·90	− 0·36782 50	+1398 80	+11 73	− 0·40968 59	+1558 49	+29 40	− 0·36951 05	+1409 34	+ 61 69
·91	·32743 34	1575 53	12 25	·40102 85	1867 10	31 22	·39827 15	1876 83	67 31
·92	·27128 65	1764 51	12 76	·37370 01	2206 93	33 24	·40826 42	2411 63	73 35
·93	·19749 45	1966 25	13 28	·32430 24	2580 00	35 28	·39414 06	3019 78	79 67
·94	− ·10404 00	2181 27	13 82	·24910 47	2988 35	37 40	·34981 92	3707 60	86 38
0·95	+0·01122 72	+2410 11	+14 39	− 0·14402 35	+3434 10	+39 60	− 0·26842 18	+4481 80	+ 93 50
·96	·15059 55	2653 34	14 92	− ·00460 13	3919 45	41 93	− ·14220 64	5349 50	100 97
·97	·31649 72	2911 49	15 54	+ ·17401 54	4446 73	44 29	+ ·03750 40	6318 17	108 89
·98	·51151 38	3185 18	16 07	·39709 94	5018 30	46 79	·28039 61	7395 73	117 22
0·99	0·73838 22	3474 94	16 74	0·67036 64	5636 66	49 31	0·59724 55	8590 51	125 97
1·00	+1·00000 00	+3781 44	+17 36	+1·00000 00	+6304 33	+52 00	+1·00000 00	+9911 26	+135 23

* Where no fourth differences are given the second differences have been modified; see page A 4.

x	$P_7(x)$	$*\delta^2$	$P_8(x)$	$*\delta^2$	δ^4	$P_9(x)$	$*\delta^2$	δ^4
1·00	1·00000 0	3778 2	1·00000 0	6304 3	52 1	1·00000 0	9911 3	135 1
·01	1·29943 2	4102 0	1·39267 7	7024 0	54 7	1·50186 7	11367 3	144 7
·02	1·63991 7	4443 5	1·85559 4	7798 4	57 6	2·11740 7	12968 0	155 5
·03	2·02487 2	4804 0	2·39649 5	8630 4	60 6	2·86262 7	14724 2	165 5
·04	2·45790 2	5183 3	3·02370 0	9523 0	63 4	3·75508 9	16645 9	177 1
1·05	2·94280 2	5582 8	3·74613 5	10479 0	67 0	4·81401 0	18744 7	188 9
·06	3·48356 8	6003 0	4·57336 0	11502 0	69 7	6·06037 8	21032 4	200 9
·07	4·08440 3	6444 2	5·51560 5	12594 7	73 6	7·51707 0	23521 0	214 2
·08	4·74972 1	6907 6	6·58379 7	13761 0	76 6	9·20897 2	26223 8	227 3
·09	5·48415 7	7393 9	7·78959 9	15003 9	80 5	11·16311 2	29153 9	241 7
1·10	6·29257 5	7903 5	9·14544 0	16327 3	84 1	13·40879 1	32325 7	256 5
·11	7·18007 3	8437 4	10·66455 4	17734 8	88 1	15·97772 7	35754 0	271 4
·12	8·15199 1	8996 3	12·36101 6	19230 4	91 4	18·90420 3	39453 7	288 0
·13	9·21392 0	9581 2	14·24978 2	20817 4	96 2	22·22521 6	43441 4	304 5
·14	10·37171 0	10192 5	16·34672 2	22500 6	100 1	25·98064 3	47733 6	322 0
1·15	11·63147 6	10831 5	18·66866 8	24283 9	104 2	30·21341	52284	
·16	12·99960 9	11498 4	21·23345 3	26171 4	108 8	34·96965	57236	
·17	14·48278 0	12194 8	24·05995 2	28167 7	113 4	40·29891	62546	
·18	16·08795 4	12921 0	27·16812 8	30277 4	118 1	46·25433	68236	
·19	17·82239 5	13678 0	30·57907 8	32505 2	122 8	52·89284	74324	
1·20	19·69367 5	14467 1	34·31508	34833		60·27536	80831	
·21	21·70968 6	15288 5	38·39964	37309		68·46701	87784	
·22	23·87864 4	16143 9	42·85754	39921		77·53736	95202	
·23	26·20910 4	17033 3	47·71490	42667		87·56063	1·03108	
·24	28·70996 3	17958 7	52·99920	45561		98·61593	1·11534	
1·25	31·39048	18918	58·73938	48602		110·78756	1·20494	
·26	34·26026	19922	64·96586	51796		124·16518	1·30027	
·27	37·32932	20956	71·71060	55155		138·84416	1·40152	
·28	40·60802	22034	79·00719	58675		154·92581	1·50902	
·29	44·10714	23154	86·89085	62373		172·51769	1·62309	
1·30	47·83787	24311	95·39856	66244		191·73391	1·74394	
·31	51·81179	25515	104·56906	70306		212·69540	1·87205	
·32	56·04094	26759	114·44297	74559		235·53031	2·00759	
·33	60·53777	28052	125·06283	79006		260·37426	2·15101	
·34	65·31520	29390	136·47314	83667		287·37073	2·30261	
1·35	70·38661	30770	148·72050	88533		316·6714	2·4627	
·36	75·76582	32208	161·85360	93623		348·4365	2·6320	
·37	81·46719	33686	175·92335	98943		382·8353	2·8102	
·38	87·50552	35220	190·98296	1·04493		420·0462	2·9988	
·39	93·89615	36806	207·08795	1·10291		460·2577	3·1969	
1·40	100·65494	38444	224·29631	1·16337		503·6681	3·4060	
·41	107·79827	40137	242·66852	1·22646		550·4865	3·6260	
·42	115·34307	41886	262·26768	1·29219		600·9330	3·8576	
·43	123·30683	43691	283·15954	1·36072		655·2393	4·1010	
·44	131·70761	45556	305·41264	1·43208		713·6489	4·3571	
1·45	140·56406	47480	329·09836	1·50640		776·4180	4·6260	
·46	149·89542	49464	354·29104	1·58374		843·8156	4·9084	
·47	159·72154	51512	381·06804	1·66421		916·1242	5·2050	
·48	170·06290	53624	409·50985	1·74794		993·6405	5·5161	
·49	180·94062	55799	439·70021	1·83497		1076·6757	5·8424	
1·50	192·37646	58046	471·72617	1·92540		1165·5562	6·1842	

* Where no fourth differences are given the second differences have been modified; see page A 4.

x	$P_7(x)$	$*\delta^2$	$P_8(x)$	$*\delta^2$	$P_9(x)$	$*\delta^2$
1·50	192·3765	5805	471·7262	1·9254	1165·5562	6·1842
·51	204·3929	6036	505·6782	2·0195	1260·6240	6·5428
·52	217·0130	6274	541·6503	2·1170	1362·2377	6·9182
·53	230·2606	6518	579·7401	2·2184	1470·7728	7·3112
·54	244·1602	6773	620·0490	2·3235	1586·6224	7·7223
1·55	258·7372	7030	662·6822	2·4329	1710·1978	8·1526
·56	274·0174	7300	707·7490	2·5460	1841·9294	8·6024
·57	290·0277	7573	755·3626	2·6634	1982·2672	9·0728
·58	306·7955	7859	805·6404	2·7852	2131·6817	9·5642
·59	324·3493	8148	858·7042	2·9115	2290·6644	10·0774
1·60	342·7181	8449	914·6803	3·0421	2459·7287	10·6134
·61	361·9319	8755	973·6994	3·1775	2639·4107	11·1729
·62	382·0214	9074	1035·8969	3·3179	2830·2700	11·7562
·63	403·0184	9397	1101·4132	3·4629	3032·8902	12·3655
·64	424·9553	9731	1170·3934	3·6132	3247·8806	12·9999
1·65	447·8655	1·0075	1242·9878	3·7687	3475·876	13·662
·66	471·7834	1·0429	1319·3519	3·9296	3717·538	14·352
·67	496·7443	1·0788	1399·6466	4·0958	3973·557	15·069
·68	522·7842	1·1162	1484·0382	4·2680	4244·651	15·817
·69	549·9404	1·1541	1572·6988	4·4455	4531·568	16·596
1·70	578·2509	1·1935	1665·8060	4·6293	4835·087	17·407
·71	607·7550	1·2335	1763·5436	4·8190	5156·019	18·250
·72	638·4928	1·2744	1866·1014	5·0151	5495·207	19·126
·73	670·5053	1·3171	1973·6755	5·2175	5853·528	20·040
·74	703·8350	1·3602	2086·4683	5·4266	6231·895	20·986
1·75	738·5251	1·4048	2204·6889	5·6421	6631·255	21·972
·76	774·6201	1·4499	2328·5529	5·8651	7052·594	22·994
·77	812·1653	1·4969	2458·2832	6·0944	7496·935	24·060
·78	851·2076	1·5447	2594·1093	6·3317	7965·343	25·162
·79	891·7948	1·5935	2736·2684	6·5758	8458·921	26·308
1·80	933·9758	1·6439	2885·0047	6·8277	8978·815	27·498
·81	977·8009	1·6951	3040·5702	7·0878	9526·215	28·733
·82	1023·3214	1·7480	3203·2249	7·3553	10102·356	30·012
·83	1070·5901	1·8019	3373·2364	7·6312	10708·518	31·340
·84	1119·6609	1·8569	3550·8806	7·9154	11346·029	32·715
1·85	1170·5889	1·9137	3736·4418	8·2083	12016·265	34·144
·86	1223·4308	1·9713	3930·2129	8·5097	12720·654	35·622
·87	1278·2443	2·0305	4132·4954	8·8202	13460·675	37·154
·88	1335·0886	2·0913	4343·5998	9·1400	14237·860	38·742
·89	1394·0244	2·1530	4563·8459	9·4688	15053·797	40·384
1·90	1455·1135	2·2167	4793·5626	9·8077	15910·129	42·087
·91	1518·4195	2·2813	5033·0887	10·1558	16808·559	43·849
·92	1584·0071	2·3478	5282·7725	10·5145	17750·849	45·670
·93	1651·9427	2·4154	5542·9726	10·8829	18738·821	47·558
·94	1722·2940	2·4848	5814·0576	11·2621	19774·363	49·510
1·95	1795·1304	2·5557	6096·4067	11·6521	20859·427	51·526
·96	1870·5228	2·6283	6390·4099	12·0527	21996·030	53·617
·97	1948·5437	2·7019	6696·4679	12·4647	23186·262	55·772
·98	2029·2669	2·7780	7014·9927	12·8883	24432·280	58·003
1·99	2112·7683	2·8549	7346·4079	13·3230	25736·315	60·308
2·00	2199·1250	2·9341	7691·1484	13·7703	27100·672	62·689

* These differences have been modified; see page A 4.

x	$P_7(x)$	δ^2	$P_8(x)$	$*\delta^2$	$P_9(x)$	$*\delta^2$
2·00	2199·125	2·934	7691·148	13·771	27100·67	62·68
·01	2288·416	3·015	8049·661	14·230	28527·73	65·16
·02	2380·722	3·098	8422·406	14·701	30019·96	67·68
·03	2476·126	3·181	8809·854	15·184	31579·89	70·33
·04	2574·711	3·268	9212·489	15·684	33210·16	73·02
2·05	2676·564	3·354	9630·810	16·191	34913·47	75·81
·06	2781·771	3·446	10065·325	16·720	36692·61	78·73
·07	2890·424	3·536	10516·561	17·253	38550·49	81·66
·08	3002·613	3·629	10985·053	17·808	40490·06	84·78
·09	3118·431	3·726	11471·355	18·372	42514·42	87·93
2·10	3237·975	3·821	11976·032	18·953	44626·73	91·19
·11	3361·340	3·921	12499·665	19·551	46830·25	94·59
·12	3488·626	4·022	13042·851	20·160	49128·37	98·05
·13	3619·934	4·126	13606·200	20·789	51524·56	101·63
·14	3755·368	4·231	14190·340	21·427	54022·40	105·33
2·15	3895·033	4·338	14795·911	22·091	56625·59	109·13
·16	4039·036	4·447	15423·575	22·761	59337·93	113·06
·17	4187·486	4·560	16074·004	23·458	62163·35	117·08
·18	4340·496	4·672	16747·893	24·162	65105·88	121·26
·19	4498·178	4·790	17445·948	24·894	68169·69	125·55
2·20	4660·650	4·907	18168·899	25·634	71359·07	129·94
·21	4828·029	5·029	18917·488	26·399	74678·42	134·51
·22	5000·437	5·150	19692·479	27·180	78132·30	139·16
·23	5177·995	5·276	20494·653	27·980	81725·37	144·00
·24	5360·829	5·404	21324·810	28·797	85462·46	148·93
2·25	5549·067	5·534	22183·768	29·639	89348·51	154·05
·26	5742·839	5·667	23072·368	30·495	93388·63	159·27
·27	5942·278	5·801	23991·467	31·375	97588·05	164·69
·28	6147·518	5·938	24941·945	32·275	1 01952·18	170·22
·29	6358·696	6·080	25924·702	33·195	1 06486·56	175·93
2·30	6575·954	6·221	26940·658	34·138	1 11196·90	181·79
·31	6799·433	6·367	27990·756	35·102	1 16089·06	187·82
·32	7029·279	6·514	29075·960	36·089	1 21169·07	194·03
·33	7265·639	6·666	30197·257	37·096	1 26443·14	200·39
·34	7508·665	6·819	31355·655	38·131	1 31917·63	206·94
2·35	7758·510	6·974	32552·188	39·184	1 37599·09	213·68
·36	8015·329	7·134	33787·910	40·265	1 43494·26	220·56
·37	8279·282	7·296	35063·901	41·368	1 49610·03	227·70
·38	8550·531	7·459	36381·265	42·499	1 55953·53	234·96
·39	8829·239	7·628	37741·132	43·650	1 62532·03	242·47
2·40	9115·575	7·798	39144·654	44·830	1 69353·03	250·15
·41	9409·709	7·971	40593·011	46·038	1 76424·22	258·05
·42	9711·814	8·149	42087·410	47·266	1 83753·50	266·16
·43	10022·068	8·327	43629·081	48·528	1 91348·98	274·46
·44	10340·649	8·510	45219·285	49·815	1 99218·97	283·03
2·45	10667·740	8·696	46859·309	51·126	2 07372·03	291·79
·46	11003·527	8·885	48550·465	52·472	2 15816·92	300·76
·47	11348·199	9·077	50294·098	53·844	2 24562·62	310·02
·48	11701·948	9·273	52091·580	55·245	2 33618·38	319·47
·49	12064·970	9·471	53944·312	56·674	2 42993·66	329·18
2·50	12437·463	9·674	55853·724	58·138	2 52698·17	339·16

* These differences have been modified; see page A 4.

x	$P_7(x)$	δ^2	$P_8(x)$	δ^2	$P_9(x)$	δ^2
2·50	12437·463	9·674	55853·72	58·15	2 52698·2	339·2
·51	12819·630	9·880	57821·28	59·63	2 62741·9	349·4
·52	13211·677	10·087	59848·47	61·16	2 73135·0	359·9
·53	13613·811	10·301	61936·82	62·71	2 83888·0	370·6
·54	14026·246	10·516	64087·88	64·31	2 95011·6	381·7
2·55	14449·197	10·736	66303·25	65·92	3 06516·9	392·9
·56	14882·884	10·959	68584·54	67·58	3 18415·1	404·5
·57	15327·530	11·187	70933·41	69·27	3 30717·8	416·4
·58	15783·363	11·415	73351·55	70·99	3 43436·9	428·5
·59	16250·611	11·651	75840·68	72·75	3 56584·5	440·9
2·60	16729·510	11·888	78402·56	74·54	3 70173·0	453·7
·61	17220·297	12·131	81038·98	76·38	3 84215·2	466·9
·62	17723·215	12·375	83751·78	78·24	3 98724·3	480·1
·63	18238·508	12·626	86542·82	80·15	4 13713·5	493·9
·64	18766·427	12·879	89414·01	82·09	4 29196·6	507·9
2·65	19307·225	13·136	92367·29	84·07	4 45187·6	522·4
·66	19861·159	13·399	95404·64	86·10	4 61701·0	536·9
·67	20428·492	13·663	98528·09	88·15	4 78751·3	552·2
·68	21009·488	13·933	1 01739·69	90·26	4 96353·8	567·6
·69	21604·417	14·208	1 05041·55	92·40	5 14523·9	583·3
2·70	22213·554	14·487	1 08435·81	94·59	5 33277·3	599·5
·71	22837·178	14·767	1 11924·66	96·80	5 52630·2	616·0
·72	23475·569	15·056	1 15510·31	99·09	5 72599·1	633·1
·73	24129·016	15·347	1 19195·05	101·39	5 93201·1	650·3
·74	24797·810	15·641	1 22981·18	103·76	6 14453·4	668·1
2·75	25482·245	15·944	1 26871·07	106·15	6 36373·8	686·2
·76	26182·624	16·247	1 30867·11	108·61	6 58980·4	704·8
·77	26899·250	16·557	1 34971·76	111·11	6 82291·8	723·7
·78	27632·433	16·870	1 39187·52	113·63	7 06326·9	743·2
·79	28382·486	17·190	1 43516·91	116·24	7 31105·2	763·1
2·80	29149·729	17·513	1 47962·54	118·88	7 56646·6	783·2
·81	29934·485	17·841	1 52527·05	121·56	7 82971·2	804·2
·82	30737·082	18·174	1 57213·12	124·30	8 10100·0	825·3
·83	31557·853	18·512	1 62023·49	127·08	8 38054·1	847·1
·84	32397·136	18·856	1 66960·94	129·94	8 66855·3	869·1
2·85	33255·275	19·203	1 72028·33	132·83	8 96525·6	892·0
·86	34132·617	19·556	1 77228·55	135·76	9 27087·9	915·1
·87	35029·515	19·915	1 82564·53	138·77	9 58565·3	938·7
·88	35946·328	20·277	1 88039·28	141·83	9 90981·4	963·1
·89	36883·418	20·646	1 93655·86	144·93	10 24360·6	987·6
2·90	37841·154	21·021	1 99417·37	148·09	10 58727·4	1013·1
·91	38819·911	21·399	2 05326·97	151·33	10 94107·3	1038·9
·92	39820·067	21·783	2 11387·90	154·60	11 30526·1	1065·2
·93	40842·006	22·175	2 17603·43	157·94	11 68010·1	1092·2
·94	41886·120	22·569	2 23976·90	161·34	12 06586·3	1119·7
2·95	42952·803	22·971	2 30511·71	164·79	12 46282·2	1147·8
·96	44042·457	23·378	2 37211·31	168·32	12 87125·9	1176·6
·97	45155·489	23·789	2 44079·23	171·89	13 29146·2	1206·0
·98	46292·310	24·208	2 51119·04	175·55	13 72372·5	1235·7
2·99	47453·339	24·632	2 58334·40	179·24	14 16834·5	1266·5
3·00	48639·000	25·062	2 65729·00	183·02	14 62563·0	1297·6

x	$P_7(x)$	δ^2	$P_8(x)$	δ^2	$P_9(x)$	δ^2
3·00	48639·00	25·06	2 65729·0	183·0	14 62563·0	1297·6
·01	49849·72	25·50	2 73306·6	186·9	15 09589·1	1329·6
·02	51085·94	25·94	2 81071·1	190·8	15 57944·8	1362·0
·03	52348·10	26·39	2 89026·4	194·6	16 07662·5	1395·3
·04	53636·65	26·84	2 97176·3	198·9	16 58775·5	1429·1
3·05	54952·04	27·30	3 05525·1	202·8	17 11317·6	1463·9
·06	56294·73	27·77	3 14076·7	207·1	17 65323·6	1499·0
·07	57665·19	28·24	3 22835·4	211·3	18 20828·6	1535·1
·08	59063·89	28·71	3 31805·4	215·6	18 77868·7	1572·0
·09	60491·30	29·21	3 40991·0	220·0	19 36480·8	1609·3
3·10	61947·92	29·70	3 50396·6	224·6	19 96702·2	1647·7
·11	63434·24	30·20	3 60026·8	228·9	20 58571·3	1686·7
·12	64950·76	30·69	3 69885·9	233·7	21 22127·1	1726·6
·13	66497·97	31·21	3 79978·7	238·4	21 87409·5	1767·1
·14	68076·39	31·74	3 90309·9	243·1	22 54459·0	1808·7
3·15	69686·55	32·25	4 00884·2	247·9	23 23317·2	1850·9
·16	71328·96	32·79	4 11706·4	252·9	23 94026·3	1893·9
·17	73004·16	33·32	4 22781·5	258·0	24 66629·3	1938·0
·18	74712·68	33·88	4 34114·6	262·9	25 41170·3	1982·8
·19	76455·08	34·43	4 45710·6	268·3	26 17694·1	2028·5
3·20	78231·91	34·98	4 57574·9	273·3	26 96246	2076
·21	80043·72	35·55	4 69712·5	278·9	27 76874	2122
·22	81891·08	36·13	4 82129·0	284·2	28 59624	2171
·23	83774·57	36·72	4 94829·7	289·7	29 44545	2220
·24	85694·78	37·29	5 07820·1	295·4	30 31686	2271
3·25	87652·28	37·90	5 21105·9	301·0	31 21098	2322
·26	89647·68	38·51	5 34692·7	306·8	32 12832	2374
·27	91681·59	39·11	5 48586·3	312·8	33 06940	2428
·28	93754·61	39·73	5 62792·7	318·6	34 03476	2481
·29	95867·36	40·36	5 77317·7	324·8	35 02493	2538
3·30	98020·47	41·01	5 92167·5	330·9	36 04048	2593
·31	1 00214·59	41·63	6 07348·2	337·2	37 08196	2651
·32	1 02450·34	42·30	6 22866·1	343·5	38 14995	2710
·33	1 04728·39	42·95	6 38727·5	350·0	39 24504	2769
·34	1 07049·39	43·62	6 54938·9	356·5	40 36782	2829
3·35	1 09414·01	44·30	6 71506·8	363·3	41 51889	2892
·36	1 11822·93	44·98	6 88438·0	369·8	42 69888	2955
·37	1 14276·83	45·67	7 05739·0	377·0	43 90842	3018
·38	1 16776·40	46·37	7 23417·0	383·7	45 14814	3085
·39	1 19322·34	47·09	7 41478·7	390·8	46 41871	3150
3·40	1 21915·37	47·81	7 59931·2	398·1	47 72078	3219
·41	1 24556·21	48·52	7 78781·8	405·4	49 05504	3287
·42	1 27245·57	49·26	7 98037·8	412·7	50 42217	3358
·43	1 29984·19	50·02	8 17706·5	420·2	51 82288	3429
·44	1 32772·83	50·75	8 37795·4	428·0	53 25788	3503
3·45	1 35612·22	51·54	8 58312·3	435·5	54 72791	3575
·46	1 38503·15	52·28	8 79264·7	443·5	56 23369	3653
·47	1 41446·36	53·09	9 00660·6	451·5	57 77600	3728
·48	1 44442·66	53·86	9 22508·0	459·5	59 35559	3807
·49	1 47492·82	54·67	9 44814·9	467·8	60 97325	3887
3·50	1 50597·65	55·47	9 67589·6	476·0	62 62978	3967

x	$P_7(x)$	δ^2	$P_8(x)$	δ^2	$P_9(x)$	δ^2
3·50	1 50597·65	55·47	9 67589·6	476·0	62 62978	3967
·51	1 53757·95	56·29	9 90840·3	484·6	64 32598	4050
·52	1 56974·54	57·13	10 14575·6	493·2	66 06268	4134
·53	1 60248·26	57·95	10 38804·1	501·7	67 84072	4219
·54	1 63579·93	58·80	10 63534·3	510·7	69 66095	4306
3·55	1 66970·40	59·67	10 88775·2	519·6	71 52424	4396
·56	1 70420·54	60·52	11 14535·7	528·7	73 43149	4483
·57	1 73931·20	61·41	11 40824·9	538·0	75 38357	4577
·58	1 77503·27	62·28	11 67652·1	547·2	77 38142	4670
·59	1 81137·62	63·20	11 95026·5	556·9	79 42597	4763
3·60	1 84835·17	64·09	12 22957·8	566·3	81 51815	4860
·61	1 88596·81	65·01	12 51455·4	576·1	83 65893	4960
·62	1 92423·46	65·94	12 80529·1	586·2	85 84931	5057
·63	1 96316·05	66·87	13 10189·0	596·0	88 09026	5160
·64	2 00275·51	67·84	13 40444·9	606·3	90 38281	5262
3·65	2 04302·81	68·77	13 71307·1	616·6	92 72798	5368
·66	2 08398·88	69·77	14 02785·9	627·2	95 12683	5475
·67	2 12564·72	70·73	14 34891·9	637·6	97 58043	5582
·68	2 16801·29	71·74	14 67635·5	648·5	100 08985	5694
·69	2 21109·60	72·72	15 01027·6	659·4	102 65621	5804
3·70	2 25490·63	73·75	15 35079·1	670·5	105 28061	5920
·71	2 29945·41	74·78	15 69801·1	681·7	107 96421	6036
·72	2 34474·97	75·80	16 05204·8	693·0	110 70817	6152
·73	2 39080·33	76·86	16 41301·5	704·7	113 51365	6274
·74	2 43762·55	77·92	16 78102·9	716·2	116 38187	6394
3·75	2 48522·69	78·98	17 15620·5	728·3	119 31403	6519
·76	2 53361·81	80·08	17 53866·4	740·1	122 31138	6645
·77	2 58281·01	81·16	17 92852·4	752·4	125 37518	6772
·78	2 63281·37	82·27	18 32590·8	764·7	128 50670	6903
·79	2 68364·00	83·40	18 73093·9	777·2	131 70725	7035
3·80	2 73530·03	84·51	19 14374·2	789·9	134 97815	7169
·81	2 78780·57	85·67	19 56444·4	802·9	138 32074	7305
·82	2 84116·78	86·82	19 99317·5	815·7	141 73638	7445
·83	2 89539·81	87·99	20 43006·3	829·1	145 22647	7585
·84	2 95050·83	89·16	20 87524·2	842·3	148 79241	7729
3·85	3 00651·01	90·36	21 32884·4	856·0	152 43564	7874
·86	3 06341·55	91·57	21 79100·6	869·7	156 15761	8022
·87	3 12123·66	92·77	22 26186·5	883·7	159 95980	8173
·88	3 17998·54	94·02	22 74156·1	897·6	163 84372	8324
·89	3 23967·44	95·25	23 23023·3	912·1	167 81088	8480
3·90	3 30031·59	96·52	23 72802·6	926·5	171 86284	8638
·91	3 36192·26	97·78	24 23508·4	941·1	176 00118	8798
·92	3 42450·71	99·06	24 75155·3	956·2	180 22750	8959
·93	3 48808·22	100·36	25 27758·4	971·0	184 54341	9126
·94	3 55266·09	101·68	25 81332·5	986·4	188 95058	9292
3·95	3 61825·64	102·99	26 35893·0	1001·8	193 45067	9464
·96	3 68488·18	104·34	26 91455·3	1017·4	198 04540	9636
·97	3 75255·06	105·67	27 48035·0	1033·4	202 73649	9812
·98	3 82127·61	107·06	28 05648·1	1049·4	207 52570	9990
3·99	3 89107·22	108·42	28 64310·6	1065·7	212 41481	10172
4·00	3 96195·25	109·82	29 24038·8	1081·9	217 40564	10355

x	$P_7(x)$	δ^2	$P_8(x)$	δ^2	$10^{-1}.P_9(x)$	δ^2
4·00	3 96195·2	109·8	29 24039	1082	217 4056	1036
·01	4 03393·1	111·2	29 84849	1099	222 5000	1054
·02	4 10702·2	112·6	30 46758	1116	227 6998	1073
·03	4 18123·9	114·1	31 09783	1133	233 0069	1093
·04	4 25659·7	115·6	31 73941	1150	238 4233	1112
4·05	4 33311·1	116·9	32 39249	1167	243 9509	1131
·06	4 41079·4	118·5	33 05724	1186	249 5916	1152
·07	4 48966·2	120·0	33 73385	1204	255 3475	1173
·08	4 56973·0	121·5	34 42250	1221	261 2207	1193
·09	4 65101·3	123·0	35 12336	1240	267 2132	1215
4·10	4 73352·6	124·5	35 83662	1260	273 3272	1235
·11	4 81728·4	126·2	36 56248	1277	279 5647	1257
·12	4 90230·4	127·6	37 30111	1297	285 9279	1280
·13	4 98860·0	129·3	38 05271	1316	292 4191	1302
·14	5 07618·9	130·8	38 81747	1336	299 0405	1324
4·15	5 16508·6	132·5	39 59559	1357	305 7943	1348
·16	5 25530·8	134·1	40 38728	1375	312 6829	1370
·17	5 34687·1	135·8	41 19272	1396	319 7085	1396
·18	5 43979·2	137·4	42 01212	1417	326 8737	1418
·19	5 53408·7	139·2	42 84569	1438	334 1807	1443
4·20	5 62977·4	140·7	43 69364	1459	341 6320	1468
·21	5 72686·8	142·6	44 55618	1480	349 2301	1493
·22	5 82538·8	144·2	45 43352	1502	356 9775	1519
·23	5 92535·0	146·0	46 32588	1524	364 8768	1545
·24	6 02677·2	147·8	47 23348	1546	372 9306	1570
4·25	6 12967·2	149·5	48 15654	1568	381 1414	1598
·26	6 23406·7	151·4	49 09528	1592	389 5120	1624
·27	6 33997·6	153·1	50 04994	1613	398 0450	1652
·28	6 44741·6	155·0	51 02073	1638	406 7432	1680
·29	6 55640·6	156·9	52 00790	1661	415 6094	1708
4·30	6 66696·5	158·7	53 01168	1684	424 6464	1737
·31	6 77911·1	160·6	54 03230	1709	433 8571	1765
·32	6 89286·3	162·5	55 07001	1733	443 2443	1795
·33	7 00824·0	164·4	56 12505	1757	452 8110	1825
·34	7 12526·1	166·4	57 19766	1784	462 5602	1855
4·35	7 24394·6	168·3	58 28811	1807	472 4949	1886
·36	7 36431·4	170·3	59 39663	1834	482 6182	1917
·37	7 48638·5	172·3	60 52349	1859	492 9332	1949
·38	7 61017·9	174·4	61 66894	1886	503 4431	1981
·39	7 73571·7	176·2	62 83325	1911	514 1511	2012
4·40	7 86301·7	178·5	64 01667	1939	525 0603	2047
·41	7 99210·2	180·4	65 21948	1966	536 1742	2079
·42	8 12299·1	182·5	66 44195	1993	547 4960	2114
·43	8 25570·5	184·8	67 68435	2021	559 0292	2148
·44	8 39026·7	186·7	68 94696	2048	570 7772	2181
4·45	8 52669·6	188·9	70 23005	2078	582 7433	2219
·46	8 66501·4	191·1	71 53392	2105	594 9313	2253
·47	8 80524·3	193·3	72 85884	2135	607 3446	2290
·48	8 94740·5	195·5	74 20511	2164	619 9869	2327
·49	9 09152·2	197·8	75 57302	2194	632 8619	2363
4·50	9 23761·7	199·8	76 96287	2223	645 9732	2402

x	$P_7(x)$	δ^2	$P_8(x)$	δ^2	$10^{-1}.P_9(x)$	δ^2
4.50	9 23761.7	199.8	76 96287	2223	645 9732	2402
.51	9 38571.0	202.4	78 37495	2255	659 3247	2440
.52	9 53582.7	204.4	79 80958	2284	672 9202	2478
.53	9 68798.8	206.8	81 26705	2316	686 7635	2519
.54	9 84221.7	209.2	82 74768	2346	700 8587	2557
4.55	9 99853.8	211.6	84 25177	2380	715 2096	2599
.56	10 15697.5	213.7	85 77966	2410	729 8204	2639
.57	10 31754.9	216.4	87 33165	2443	744 6951	2680
.58	10 48028.7	218.6	88 90807	2477	759 8378	2724
.59	10 64521.1	221.1	90 50926	2508	775 2529	2764
4.60	10 81234.6	223.6	92 13553	2544	790 9444	2810
.61	10 98171.7	226.0	93 78724	2575	806 9169	2852
.62	11 15334.8	228.6	95 46470	2612	823 1746	2896
.63	11 32726.5	231.1	97 16828	2645	839 7219	2942
.64	11 50349.3	233.5	98 89831	2681	856 5634	2988
4.65	11 68205.6	236.2	100 65515	2715	873 7037	3032
.66	11 86298.1	238.8	102 43914	2752	891 1472	3081
.67	12 04629.4	241.3	104 25065	2788	908 8988	3126
.68	12 23202.0	244.0	106 09004	2824	926 9630	3177
.69	12 42018.6	246.7	107 95767	2862	945 3449	3223
4.70	12 61081.9	249.3	109 85392	2899	964 0491	3273
.71	12 80394.5	252.1	111 77916	2936	983 0806	3323
.72	12 99959.2	254.7	113 73376	2974	1002 4444	3374
.73	13 19778.6	257.6	115 71810	3015	1022 1456	3425
.74	13 39855.6	260.2	117 73259	3051	1042 1893	3476
4.75	13 60192.8	263.1	119 77759	3092	1062 5806	3529
.76	13 80793.1	265.9	121 85351	3132	1083 3248	3582
.77	14 01659.3	268.7	123 96075	3172	1104 4272	3636
.78	14 22794.2	271.7	126 09971	3213	1125 8932	3691
.79	14 44200.8	274.4	128 27080	3253	1147 7283	3746
4.80	14 65881.8	277.5	130 47442	3296	1169 9380	3802
.81	14 87840.3	280.3	132 71100	3337	1192 5279	3858
.82	15 10079.1	283.4	134 98095	3380	1215 5036	3916
.83	15 32601.3	286.2	137 28470	3423	1238 8709	3975
.84	15 55409.7	289.5	139 62268	3465	1262 6357	4032
4.85	15 78507.6	292.2	141 99531	3511	1286 8037	4092
.86	16 01897.7	295.6	144 40305	3554	1311 3809	4154
.87	16 25583.4	298.5	146 84633	3600	1336 3735	4213
.88	16 49567.6	301.6	149 32561	3643	1361 7874	4277
.89	16 73853.4	304.8	151 84132	3690	1387 6290	4337
4.90	16 98444.0	308.0	154 39393	3737	1413 9043	4403
.91	17 23342.6	311.2	156 98391	3781	1440 6199	4466
.92	17 48552.4	314.3	159 61170	3831	1467 7821	4531
.93	17 74076.5	317.8	162 27780	3878	1495 3974	4597
.94	17 99918.4	320.8	164 98268	3924	1523 4724	4664
4.95	18 26081.1	324.2	167 72680	3975	1552 0138	4731
.96	18 52568.0	327.6	170 51067	4023	1581 0283	4799
.97	18 79382.5	330.9	173 33477	4073	1610 5227	4869
.98	19 06527.9	334.3	176 19960	4123	1640 5040	4938
4.99	19 34007.6	337.7	179 10566	4173	1670 9791	5010
5.00	19 61825.0	341.1	182 05345	4225	1701 9552	5082

x	$P_7(x)$	δ^2	$P_8(x)$	δ^2	$10^{-1}.P_9(x)$	δ^2
5·00	19 61825·0	341·1	182 05345	4225	1701 9552	5082
·01	19 89983·5	344·7	185 04349	4276	1733 4395	5153
·02	20 18486·7	348·1	188 07629	4329	1765 4391	5227
·03	20 47338·0	351·6	191 15238	4380	1797 9614	5302
·04	20 76540·9	355·2	194 27227	4436	1831 0139	5377
5·05	21 06099·0	358·8	197 43652	4487	1864 6041	5453
·06	21 36015·9	362·3	200 64564	4543	1898 7396	5531
·07	21 66295·1	366·1	203 90019	4597	1933 4282	5607
·08	21 96940·4	369·7	207 20071	4652	1968 6775	5688
·09	22 27955·4	373·3	210 54775	4710	2004 4956	5768
5·10	22 59343·7	377·2	213 94189	4764	2040 8905	5847
·11	22 91109·2	380·8	217 38367	4823	2077 8701	5931
·12	23 23255·5	384·7	220 87368	4880	2115 4428	6012
·13	23 55786·5	388·5	224 41249	4938	2153 6167	6097
·14	23 88706·0	392·3	228 00068	4997	2192 4003	6182
5·15	24 22017·8	396·1	231 63884	5056	2231 8021	6267
·16	24 55725·7	400·2	235 32756	5116	2271 8306	6354
·17	24 89833·8	404·0	239 06744	5177	2312 4945	6443
·18	25 24345·9	408·1	242 85909	5238	2353 8027	6530
·19	25 59266·1	411·9	246 70312	5300	2395 7639	6621
5·20	25 94598·2	416·1	250 60015	5361	2438 3872	6713
·21	26 30346·4	420·1	254 55079	5426	2481 6818	6803
·22	26 66514·7	424·2	258 55569	5488	2525 6567	6898
·23	27 03107·2	428·3	262 61547	5553	2570 3214	6992
·24	27 40128·0	432·5	266 73078	5617	2615 6853	7086
5·25	27 77581·3	436·6	270 90226	5684	2661 7578	7184
·26	28 15471·2	441·0	275 13058	5748	2708 5487	7281
·27	28 53802·1	445·1	279 41638	5816	2756 0677	7379
·28	28 92578·1	449·4	283 76034	5883	2804 3246	7481
·29	29 31803·5	453·8	288 16313	5951	2853 3296	7579
5·30	29 71482·7	458·1	292 62543	6020	2903 0925	7684
·31	30 11620·0	462·5	297 14793	6089	2953 6238	7785
·32	30 52219·8	467·0	301 73132	6159	3004 9336	7892
·33	30 93286·6	471·3	306 37630	6229	3057 0326	7996
·34	31 34824·7	475·8	311 08357	6301	3109 9312	8103
5·35	31 76838·6	480·4	315 85385	6373	3163 6401	8212
·36	32 19332·9	485·0	320 68786	6446	3218 1702	8320
·37	32 62312·2	489·5	325 58633	6519	3273 5323	8433
·38	33 05781·0	494·1	330 54999	6592	3329 7377	8543
·39	33 49743·9	498·8	335 57957	6668	3386 7974	8657
5·40	33 94205·6	503·6	340 67583	6743	3444 7228	8771
·41	34 39170·9	508·1	345 83952	6820	3503 5253	8887
·42	34 84644·3	513·1	351 07141	6896	3563 2165	9004
·43	35 30630·8	517·8	356 37226	6973	3623 8081	9122
·44	35 77135·1	522·6	361 74284	7052	3685 3119	9242
5·45	36 24162·0	527·5	367 18394	7132	3747 7399	9364
·46	36 71716·4	532·5	372 69636	7210	3811 1043	9485
·47	37 19803·3	537·3	378 28088	7291	3875 4172	9609
·48	37 68427·5	542·3	383 93831	7374	3940 6910	9735
·49	38 17594·0	547·4	389 66948	7453	4006 9383	9861
5·50	38 67307·9	552·4	395 47518	7539	4074 1717	9988

x	$P_7(x)$	δ^2	$10^{-1} . P_8(x)$	δ^2	$10^{-2} . P_9(x)$	δ^2
5·50	38 67308	552	395 4752	754	4074 172	998
·51	39 17574	558	401 3563	762	4142 404	1 012
·52	39 68398	563	407 3136	770	4211 648	1 025
·53	40 19785	567	413 3479	779	4281 917	1 038
·54	40 71739	573	419 4601	788	4353 224	1 052
5·55	41 24266	579	425 6511	797	4425 583	1 064
·56	41 77372	583	431 9218	804	4499 006	1 080
·57	42 31061	589	438 2729	814	4573 509	1 091
·58	42 85339	594	444 7054	823	4649 103	1 108
·59	43 40211	600	451 2202	832	4725 805	1 119
5·60	43 95683	604	457 8182	840	4803 626	1 136
·61	44 51759	612	464 5002	851	4882 583	1 149
·62	45 08447	615	471 2673	858	4962 689	1 164
·63	45 65750	622	478 1202	870	5043 959	1 178
·64	46 23675	627	485 0601	876	5126 407	1 194
5·65	46 82227	633	492 0876	889	5210 049	1 208
·66	47 41412	639	499 2040	896	5294 899	1 224
·67	48 01236	643	506 4100	907	5380 973	1 240
·68	48 61703	651	513 7067	916	5468 287	1 254
·69	49 22821	656	521 0950	926	5556 855	1 270
5·70	49 84595	661	528 5759	936	5646 693	1 287
·71	50 47030	667	536 1504	947	5737 818	1 302
·72	51 10132	674	543 8196	955	5830 245	1 319
·73	51 73908	679	551 5843	967	5923 991	1 335
·74	52 38363	686	559 4457	977	6019 072	1 351
5·75	53 03504	691	567 4048	986	6115 504	1 368
·76	53 69336	698	575 4625	998	6213 304	1 386
·77	54 35866	703	583 6200	1007	6312 490	1 403
·78	55 03099	710	591 8782	1020	6413 079	1 419
·79	55 71042	717	600 2384	1028	6515 087	1 437
5·80	56 39702	721	608 7014	1041	6618 532	1 454
·81	57 09083	729	617 2685	1051	6723 431	1 474
·82	57 79193	735	625 9407	1061	6829 804	1 490
·83	58 50038	742	634 7190	1074	6937 667	1 509
·84	59 21625	747	643 6047	1084	7047 039	1 527
5·85	59 93959	755	652 5988	1096	7157 938	1 546
·86	60 67048	760	661 7025	1107	7270 383	1 564
·87	61 40897	768	670 9169	1119	7384 392	1 584
·88	62 15514	773	680 2432	1130	7499 985	1 603
·89	62 90904	782	689 6825	1141	7617 181	1 622
5·90	63 67076	786	699 2359	1154	7735 999	1 642
·91	64 44034	795	708 9047	1166	7856 459	1 662
·92	65 21787	801	718 6901	1177	7978 581	1 681
·93	66 00341	808	728 5932	1190	8102 384	1 701
·94	66 79703	814	738 6153	1202	8227 888	1 723
5·95	67 59879	822	748 7576	1213	8355 115	1 743
·96	68 40877	829	759 0212	1228	8484 085	1 763
·97	69 22704	835	769 4076	1238	8614 818	1 784
·98	70 05366	843	779 9178	1252	8747 335	1 806
5·99	70 88871	849	790 5532	1265	8881 658	1 828
6·00	71 73225	858	801 3151	1277	9017 809	1 848

x	$P_{10}(x)$	$*\delta^2$	$P_{11}(x)$	$*\delta^2$	$P_{12}(x)$	$*\delta^2$
0·00	− 0·24609 38	+ 270 99	− 0·00000 00	+ 0	+ 0·22558 59	− 352 42
·01	·24474 14	269 57	·02701 17	35 17	·22382 86	349 82
·02	·24069 84	265 35	·05367 25	69 89	·21858 28	341 92
·03	·23400 69	258 40	·07963 59	103 83	·20992 72	328 88
·04	·22473 64	248 72	·10456 34	136 46	·19799 18	310 93
0·05	− 0·21298 35	+ 236 40	− 0·12812 93	+ 167 47	+ 0·18295 58	− 288 22
·06	·19887 11	221 63	·15002 42	196 45	·16504 57	261 12
·07	·18254 68	204 46	·16995 89	223 08	·14453 18	230 02
·08	·16418 20	185 07	·18766 78	246 95	·12172 44	195 24
·09	·14397 02	163 68	·20291 27	267 82	·09697 02	157 42
0·10	− 0·12212 50	+ 140 42	− 0·21548 54	+ 285 44	+ 0·07064 66	− 116 98
·11	·09887 86	115 54	·22521 03	299 44	·04315 69	74 51
·12	·07447 93	89 31	·23194 76	309 77	+ ·01492 47	− 30 64
·13	·04918 90	61 91	·23559 44	316 19	− ·01361 24	+ 14 05
·14	− ·02328 12	33 62	·23608 67	318 59	·04200 88	58 84
0·15	+ 0·00296 18	+ 4 78	− 0·23340 06	+ 316 89	− 0·06981 77	+ 103 15
·16	·02925 20	− 24 40	·22755 31	311 11	·09659 73	146 26
·17	·05529 81	53 62	·21860 20	301 22	·12191 77	187 53
·18	·08080 85	82 58	·20664 61	287 27	·14536 73	226 35
·19	·10549 42	110 95	·19182 46	269 48	·16655 91	262 08
0·20	+ 0·12907 20	− 138 45	− 0·17431 53	+ 247 84	− 0·18513 69	+ 294 13
·21	·15126 74	164 79	·15433 40	222 75	·20078 12	321 94
·22	·17181 75	189 73	·13213 13	194 33	·21321 47	345 09
·23	·19047 36	212 84	·10799 08	162 89	·22220 68	363 05
·24	·20700 49	234 02	·08222 61	128 90	·22757 85	375 52
0·25	+ 0·22120 02	− 252 87	− 0·05517 66	+ 92 56	− 0·22920 57	+ 382 15
·26	·23287 14	269 25	− ·02720 48	54 45	·22702 25	382 73
·27	·24185 52	282 82	+ ·00130 89	+ 14 91	·22102 33	377 14
·28	·24801 62	293 49	·02997 01	− 25 51	·21126 42	365 28
·29	·25124 81	300 98	·05837 55	66 33	·19786 37	347 21
0·30	+ 0·25147 63	− 305 17	+ 0·08611 79	− 106 99	− 0·18100 22	+ 323 10
·31	·24865 91	305 95	·11279 16	147 01	·16092 05	293 10
·32	·24278 89	303 16	·13799 75	185 76	·13791 80	257 55
·33	·23389 37	296 79	·16134 91	222 73	·11234 93	216 95
·34	·22203 73	286 76	·18247 77	257 34	·08461 96	171 70
0·35	+ 0·20732 00	− 273 10	+ 0·20103 82	− 289 06	− 0·05518 02	+ 122 47
·36	·18987 83	255 83	·21671 44	317 36	− ·02452 22	69 92
·37	·16988 48	235 04	·22922 42	341 72	+ ·00683 04	+ 14 85
·38	·14754 72	210 82	·23832 49	361 66	·03832 84	− 41 94
·39	·12310 73	183 39	·24381 79	376 74	·06940 56	99 49
0·40	+ 0·09683 91	− 152 90	+ 0·24555 31	− 386 63	+ 0·09948 82	− 156 89
·41	·06904 71	119 78	·24343 24	390 84	·12800 41	213 10
·42	·04006 39	83 58	·23741 40	389 25	·15439 30	267 13
·43	+ ·01024 69	45 77	·22751 43	381 54	·17811 66	317 86
·44	− ·02002 45	− 5 97	·21381 06	367 61	·19866 94	364 36
0·45	− 0·05035 30	+ 35 26	+ 0·19644 24	− 347 37	+ 0·21558 85	− 405 46
·46	·08032 72	77 41	·17561 20	320 87	·22846 46	440 28
·47	·10952 64	120 07	·15158 42	288 21	·23695 13	467 78
·48	·13752 51	162 62	·12468 52	249 61	·24077 51	487 18
·49	·16389 87	204 59	·09530 04	205 37	·23974 35	497 65
0·50	− 0·18822 86	+ 245 35	+ 0·06387 14	− 155 94	+ 0·23375 30	− 498 54

* These differences have been modified: see page A 4.

x	$P_{10}(x)$	$*\delta^2$	$*\delta^4$	$P_{11}(x)$	$*\delta^2$	$*\delta^4$	$P_{12}(x)$	δ^2	$*\delta^4$
0·50	− 0·1882 86 +	245 35		+ 0·0638 7 14 −	155 94		+ 0·2337 5 30 −	496 68 +	10 17
·51	·2101 0 83	284 33		+ ·0308 9 16	101 81		·2227 9 57	487 35	10 42
·52	·2291 4 92	320 84		− ·0030 9 89 −	43 63		·2069 6 49	467 60	10 78
·53	·2449 8 73	354 30		·0375 1 96 +	17 89		·1864 5 81	437 14	10 70
·54	·2572 8 92	384 07		·0717 5 70	81 81		·1615 7 99	396 01	10 54
0·55	− 0·2657 5 85 +	409 42		− 0·1051 7 35 +	147 29		+ 0·1327 4 16 −	344 39 +	10 13
·56	·2701 4 28	429 77		·1371 1 63	213 19		·1004 5 94	282 69	9 46
·57	·2702 3 97	444 51		·1669 2 85	278 41		·0653 5 03	211 58	8 53
·58	·2659 0 30	442 96		·1939 6 02	341 70		+ ·0281 2 54	131 98	7 45
·59	·2570 4 92	454 63		·2175 8 09	401 75		− ·0104 1 93 −	45 01	5 90
0·60	− 0·2436 6 27 +	448 89		− 0·2371 9 27 +	456 03 −	5 95	− 0·0494 1 41 +	47 83 +	4 30
·61	·2258 0 16	435 37		·2522 4 42	505 18	7 55	·0879 3 06	144 90	2 31
·62	·2036 0 19	413 63		·2622 4 39	546 81	8 91	·1249 9 81	244 23 +	9
·63	·1772 8 16	383 33		·2667 7 55	579 52	10 41	·1596 2 33	343 60 −	2 36
·64	·1471 4 41	344 23		·2655 1 19	601 84	11 76	·1908 1 25	440 57	4 98
0·65	− 0·1135 8 05 +	296 31		− 0·2582 2 99 +	612 41 −	13 06	− 0·2175 9 60 +	532 51 −	7 89
·66	·0770 7 01	239 48		·2448 2 38	609 93	14 36	·2390 5 44	616 55	10 79
·67	− ·0381 8 08	173 98		·2253 1 84	593 14	15 34	·2543 4 73	689 76	13 96
·68	+ ·0024 3 30	100 12		·1998 8 16	561 04	16 22	·2627 4 26	749 02	17 01
·69	·0440 3 37 +	18 44		·1688 3 44	512 77	16 84	·2636 4 77	791 27	20 09
0·70	+ 0·0858 0 58 −	70 37		− 0·1326 5 95 +	447 72 −	17 21	− 0·2566 4 01 +	813 45 −	23 06
·71	·1268 6 31	165 25		·0920 0 74	365 55	17 10	·2414 9 80	812 62	25 81
·72	·1662 5 89	265 10		·0476 9 98	266 35	16 70	·2182 2 97	786 06	28 12
·73	·2029 9 76	368 28		− ·0007 2 87	150 56	15 70	·1871 0 08	731 46	30 10
·74	·2360 5 07	473 05		+ ·0477 4 80 +	19 17	14 27	·1486 5 73	646 90	31 39
0·75	+ 0·2643 7 45 −	576 64 +	3 12	+ 0·0964 1 64 −	126 34 −	12 05	− 0·1037 4 48 +	531 12 −	31 89
·76	·2869 3 19	677 14	6 11	·1438 2 14	283 75	9 13	− ·0535 2 11	383 66	31 37
·77	·3027 1 79	771 46	9 37	·1883 8 89	450 13	5 46	+ ·0005 3 92 +	205 07	29 73
·78	·3107 8 93	856 28	13 35	·2284 5 51	621 76 −	71	·0566 5 02 −	2 94	26 58
·79	·3102 9 79	927 67	17 64	·2623 0 37	793 89 +	5 01	·1127 3 18	237 18	21 74
0·80	+ 0·3005 2 98 −	981 30 +	22 57	+ 0·2882 1 34 −	960 77 +	11 91	+ 0·1664 4 16 −	492 75 −	14 96
·81	·2809 4 87	1012 24	28 07	·3045 1 54	1115 47	20 13	·2152 2 39	762 78 −	5 73
·82	·2512 4 52	1014 98	34 23	·3096 6 27	1249 74	29 83	·2563 7 84	1037 99 +	6 10
·83	·2113 9 19	983 36	41 00	·3023 1 26	1353 86	41 07	·2871 5 30	1306 46	20 98
·84	·1617 0 50	910 58	48 60	·2814 2 39	1416 55	54 04	·3048 6 30	1553 20	39 55
0·85	+ 0·1029 1 23 −	789 05 +	56 90	+ 0·2463 6 97 −	1424 79 +	69 04	+ 0·3070 4 10 −	1759 59 +	61 86
·86	+ ·0362 2 91	610 45	65 99	·1970 6 76	1363 57	86 04	·2916 2 31	1903 16	88 87
·87	− ·0365 5 86	365 66	76 13	·1341 2 98	1215 84	105 30	·2571 7 36	1956 82	120 86
·88	·1130 0 29 −	44 57	86 98	+ ·0590 3 36	962 27	127 25	·2031 5 59	1888 45	158 48
·89	·1898 9 29 +	363 73	99 03	− ·0256 8 53	580 92	151 64	·1302 5 37	1660 28	202 48
0·90	− 0·2631 4 56 +	871 26 +	111 96	− 0·1162 1 34 −	47 30 +	179 14	+ 0·0407 4 87 −	1228 18 +	253 47
·91	·3276 8 57	1491 00	126 20	·2072 1 45 +	666 12	209 84	− ·0610 3 81 −	540 99	312 29
·92	·3773 1 58	2237 16	141 38	·2915 5 44	1590 07	243 79	·1682 3 48 +	460 27	379 65
·93	·4045 7 43	3124 99	158 04	·3599 9 36	2758 60	281 67	·2708 2 88	1843 16	456 62
·94	·4005 8 29	4171 12	175 90	·4008 4 68	4209 60	323 35	·3549 9 12	3684 83	544 01
0·95	− 0·3548 8 03 +	5393 44 +	195 16	− 0·3996 0 40 +	5984 84 +	369 37	− 0·4023 0 53 +	6072 88 +	642 81
·96	·2552 4 33	6811 23	215 95	·3385 1 28	8130 41	420 06	·3888 9 06	9106 36	754 28
·97	− ·0874 9 40	8445 28	238 20	− ·1961 1 75	10697 05	475 59	·2844 1 23	12896 97	879 49
·98	+ ·1647 0 81	10317 88	262 19	+ ·0532 4 83	13740 37	536 41	− ·0509 6 43	17570 18	1019 70
0·99	·5200 8 90	12453 01	287 75	·4400 1 78	17321 27	602 90	+ ·3581 8 55	23266 47	1176 20
1·00	+ 1·0000 0 00	+ 14876 28	+ 315 26	+ 1·0000 0 00	+ 21506 32	+ 675 44	+ 1·0000 0 00	+ 30142 66	+ 1350 61

* The highest differences shown, whether δ^2 or δ^4, have been modified; see page A4.

x	$P_{10}(x)$	$*\delta^2$	δ^4	$P_{11}(x)$	$*\delta^2$	δ^4	$P_{12}(x)$	δ^2	δ^4
1·00	1·00000 0	14876 3	315 5	1·00000	21507	674	1·00000	30143	1354
·01	1·62867 4	17615 1	345 3	1·77505	26367	760	1·94324	38374	1547
·02	2·43349 9	20699 2	376 0	2·81377	31987	837	3·27022	48152	1766
·03	3·44531 6	24159 3	409 9	4·17236	38444	938	5·07872	59696	1999
·04	4·69872 6	28029 3	445 3	5·91539	45839	1036	7·48418	73239	2264
1·05	6·23242 9	32344 6	483 3	8·11681	54270	1142	10·62203	89046	2552
·06	8·08957 8	37143 2	523 5	10·86093	63843	1267	14·65034	1·07405	2868
·07	10·31815 9	42465 3	566 4	14·24348	74683	1387	19·75270	1·28632	3217
·08	12·97139 3	48353 8	611 7	18·37286	86910	1530	26·14138	1·53076	3596
·09	16·10816 5	54854 0	659 8	23·37134	1·00667	1673	34·06082	1·81116	4015
1·10	19·79347 7	62014 0	710 8	29·37649	1·16097	1836	43·79142	2·13171	4466
·11	24·09892 9	69884 8	765 0	36·54261	1·33363	2000	55·65373	2·49692	4966
·12	29·10322 9	78520 6	821 7	45·04236	1·52629	2192	70·01296	2·91179	5497
·13	34·89273 5	87978 1	882 5	55·06840	1·74087	2377	87·28398	3·38163	6090
·14	41·56202 2	98318 1	945 5	66·83531	1·97922	2596	107·93663	3·91237	6721
1·15	49·21449	1·09603	1015	80·58144	2·24353	2811	132·50165	4·51032	7413
·16	57·96299	1·21903	1082	96·57110	2·53595	3059	161·57699	5·18240	8154
·17	67·93052	1·35285	1160	115·09671	2·85896	3306	195·83473	5·93602	8965
·18	79·25090	1·49827	1236	136·48128	3·21503	3583	236·02849	6·77929	9832
·19	92·06955	1·65605	1321	161·08088	3·60693	3872	283·00154	7·72088	10775
1·20	106·54425	1·82704	1406	189·28741	4·03755	4179	337·6955	8·7701	1183
·21	122·84599	2·01209	1499	221·53149	4·50996	4510	401·1596	9·9375	1284
·22	141·15982	2·21213	1595	258·28553	5·02747	4853	474·5612	11·2333	1407
·23	161·68578	2·42812	1696	300·06704	5·59351	5231	559·1961	12·6698	1530
·24	184·63986	2·66107	1798	347·44206	6·21186	5618	656·5008	14·2593	1666
1·25	210·25501	2·91200	1917	401·0289	6·8865	599	768·0648	16·0154	1812
·26	238·78216	3·18210	2024	461·5022	7·6212	651	895·6442	17·9527	1966
·27	270·49141	3·47244	2152	529·5967	8·4210	693	1041·1763	20·0866	2135
·28	305·67310	3·78430	2281	606·1122	9·2901	744	1206·7950	22·4340	2308
·29	344·63909	4·11897	2406	691·9178	10·2336	797	1394·8477	25·0122	2504
1·30	387·7240	4·4731		787·9570	11·2568	851	1607·9126	27·8408	2706
·31	435·2867	4·8571		895·2530	12·3651	911	1848·8183	30·9400	2921
·32	487·7114	5·2678		1014·9141	13·5645	971	2120·6640	34·3313	3155
·33	545·4092	5·7075		1148·1397	14·8610	1036	2426·8410	38·0381	3399
·34	608·8200	6·1771		1296·2263	16·2611	1107	2771·0561	42·0848	3666
1·35	678·4137	6·6781		1460·5740	17·7719	1177	3157·356	46·498	395
·36	754·6917	7·2130		1642·6936	19·4004	1255	3590·154	51·306	423
·37	838·1892	7·7829		1844·2136	21·1544	1331	4074·258	56·537	456
·38	929·4765	8·3905		2066·8880	23·0415	1422	4614·899	62·224	492
·39	1029·1614	9·0364		2312·6039	25·0708	1505	5217·764	68·403	521
1·40	1137·8904	9·7242		2583·391	27·220		5889·032	75·103	566
·41	1256·3516	10·4552		2881·428	29·560		6635·403	82·369	601
·42	1385·2764	11·2315		3209·056	32·065		7464·143	90·236	648
·43	1525·4416	12·0560		3568·783	34·756		8383·119	98·751	688
·44	1677·6721	12·9308		3963·301	37·634		9400·846	107·954	741
1·45	1842·8431	13·8582		4395·490	40·715		10526·527	117·898	786
·46	2021·8825	14·8411		4868·433	44·009		11770·106	128·628	846
·47	2215·7737	15·8822		5385·427	47·529		13142·313	140·204	896
·48	2425·5583	16·9842		5949·995	51·293		14654·724	152·676	960
·49	2652·3388	18·1498		6565·902	55·306		16319·811	166·108	1·019
1·50	2897·2815	19·3829		7237·164	59·588		18151·006	180·559	1·090

* Where no fourth differences are given the second differences have been modified; see page A 4.

x	$P_{10}(x)$	*δ^2	$P_{11}(x)$	*δ^2	$P_{12}(x)$	*δ^2
1·50	2897·282	19·382	7237·164	59·588	18151·01	180·35
·51	3161·620	20·686	7968·066	64·152	20162·76	195·89
·52	3446·657	22·062	8763·175	69·014	22370·61	212·59
·53	3753·770	23·516	9627·356	74·192	24791·27	230·46
·54	4084·413	25·046	10565·790	79·701	27442·64	249·73
1·55	4440·118	26·667	11583·989	85·558	30343·99	270·33
·56	4822·505	28·368	12687·814	91·787	33515·94	292·41
·57	5233·277	30·165	13883·497	98·399	36980·59	316·07
·58	5674·232	32·059	15177·655	105·426	40761·62	341·41
·59	6147·264	34·047	16577·317	112·875	44884·38	368·50
1·60	6654·363	36·146	18089·938	120·781	49375·98	397·43
·61	7197·628	38·350	19723·427	129·160	54265·38	428·40
·62	7779·264	40·667	21486·168	138·040	59583·56	461·42
·63	8401·589	43·106	23387·045	147·442	65363·57	496·68
·64	9067·042	45·663	25435·465	157·394	71640·69	534·28
1·65	9778·182	48·353	27641·386	167·927	78452·55	574·38
·66	10537·699	51·173	30015·345	179·063	85839·27	617·08
·67	11348·415	54·133	32568·484	190·838	93843·58	662·57
·68	12213·291	57·242	35312·583	203·275	1 02511·00	710·99
·69	13135·436	60·496	38260·086	216·416	1 11889·98	762·48
1·70	14118·106	63·912	41424·14	230·29	1 22032·05	817·28
·71	15164·717	67·485	44818·62	244·92	1 32992·03	875·45
·72	16278·845	71·234	48458·17	260·35	1 44828·14	937·33
·73	17464·239	75·155	52358·23	276·65	1 57602·28	1002·96
·74	18724·822	79·262	56535·10	293·79	1 71380·13	1072·66
1·75	20064·702	83·558	61005·93	311·89	1 86231·42	1146·60
·76	21488·176	88·051	65788·82	330·91	2 02230·13	1225·00
·77	22999·739	92·751	70902·81	350·97	2 19454·70	1308·08
·78	24604·092	97·662	76367·96	372·03	2 37988·26	1396·12
·79	26306·148	102·797	82205·35	394·27	2 57918·90	1489·37
1·80	28111·04	108·17	88437·21	417·56	2 79339·9	1588·0
·81	30024·14	113·77	95086·86	442·13	3 02350·0	1692·6
·82	32051·05	119·58	1 02178·86	467·88	3 27053·8	1802·9
·83	34197·60	125·73	1 09738·98	494·97	3 53561·7	1919·8
·84	36469·92	132·08	1 17794·32	523·42	3 81990·6	2043·2
1·85	38874·37	138·73	1 26373·34	553·27	4 12464·0	2173·6
·86	41417·60	145·64	1 35505·90	584·59	4 45112·4	2311·5
·87	44106·53	152·89	1 45223·34	617·48	4 80073·7	2456·8
·88	46948·40	160·38	1 55558·55	651·95	5 17493·3	2610·4
·89	49950·72	168·27	1 66546·02	688·10	5 57524·8	2772·3
1·90	53121·36	176·41	1 78221·91	725·97	6 00330·2	2943·0
·91	56468·48	184·94	1 90624·11	765·69	6 46080·3	3123·2
·92	60000·60	193·79	2 03792·34	807·26	6 94955·3	3312·6
·93	63726·58	203·02	2 17768·19	850·82	7 47144·8	3512·7
·94	67655·65	212·62	2 32595·23	896·40	8 02848·9	3723·2
1·95	71797·41	222·61	2 48319·06	944·10	8 62278·2	3944·6
·96	76161·85	232·99	2 64987·40	994·01	9 25654·3	4178·0
·97	80759·36	243·79	2 82650·18	1046·26	9 93210·6	4423·2
·98	85600·74	255·04	3 01359·65	1100·83	10 65192·5	4681·5
1·99	90697·24	266·69	3 21170·41	1157·90	11 41858·3	4952·9
2·00	96060·52	278·84	3 42139·55	1217·60	12 23479·5	5238·0

* These differences have been modified; see page A 4.

x	$P_{10}(x)$	$*\delta^2$	$P_{11}(x)$	$*\delta^2$	$P_{12}(x)$	$*\delta^2$
2·00	96060·52	278·84	3 42139·6	1217·5	12 23479	5238
·01	1 01702·72	291·43	3 64326·8	1279·9	13 10341	5538
·02	1 07636·44	304·53	3 87794·4	1345·1	14 02744	5852
·03	1 13874·78	318·13	4 12607·6	1412·9	15 01002	6184
·04	1 20431·34	332·23	4 38834·3	1484·1	16 05447	6529
2·05	1 27320·23	346·88	4 66545·6	1558·0	17 16425	6896
·06	1 34556·10	362·09	4 95815·5	1635·5	18 34302	7278
·07	1 42154·16	377·85	5 26721·5	1716·1	19 59460	7677
·08	1 50130·18	394·23	5 59344·2	1800·1	20 92299	8101
·09	1 58500·54	411·19	5 93767·7	1887·9	22 33242	8540
2·10	1 67282·21	428·79	6 30079·8	1979·5	23 82729	9003
·11	1 76492·79	447·02	6 68372·1	2074·6	25 41223	9487
·12	1 86150·52	465·95	7 08739·8	2173·9	27 09208	9995
·13	1 96274·32	485·53	7 51282·2	2277·7	28 87192	10525
·14	2 06883·78	505·84	7 96103·0	2385·3	30 75706	11082
2·15	2 17999·21	526·87	8 43309·9	2497·4	32 75307	11665
·16	2 29641·65	548·65	8 93015·1	2614·3	34 86578	12273
·17	2 41832·88	571·20	9 45335·5	2736·0	37 10128	12913
·18	2 54595·46	594·55	10 00392·8	2862·6	39 46596	13578
·19	2 67952·74	618·73	10 58313·6	2994·1	41 96648	14277
2·20	2 81928·9	643·7	11 19229·5	3131·1	44 60983	15006
·21	2 96549·0	669·5	11 83277·5	3273·7	47 40330	15767
·22	3 11838·8	696·5	12 50600·2	3421·4	50 35451	16567
·23	3 27825·2	724·1	13 21345·5	3575·8	53 47145	17396
·24	3 44535·9	752·7	13 95667·6	3735·5	56 76242	18266
2·25	3 61999·5	782·3	14 73726·4	3901·9	60 23612	19176
·26	3 80245·6	813·2	15 55688·3	4074·7	63 90165	20121
·27	3 99305·0	844·4	16 41726·1	4253·9	67 76847	21113
·28	4 19209·1	877·4	17 32019·1	4440·4	71 84649	22142
·29	4 39990·8	911·2	18 26753·8	4634·0	76 14602	23222
2·30	4 61683·9	945·9	19 26123·8	4834·5	80 67785	24345
·31	4 84323·2	982·2	20 30329·8	5043·1	85 45322	25517
·32	5 07944·9	1019·3	21 39580·4	5259·6	90 48385	26738
·33	5 32586·2	1058·0	22 54092·1	5483·9	95 78196	28012
·34	5 58285·7	1097·5	23 74089·3	5716·8	101 36029	29340
2·35	5 85083·0	1138·5	24 99804·9	5958·4	107 23212	30723
·36	6 13019·1	1181·1	26 31480·5	6208·7	113 41129	32165
·37	6 42136·5	1224·5	27 69366·5	6468·6	119 91222	33667
·38	6 72478·7	1269·8	29 13722·8	6737·6	126 74993	35230
·39	7 04090·9	1316·2	30 64818·5	7016·5	133 94006	36858
2·40	7 37019·6	1364·2	32 22932·6	7305·8	141 49889	38552
·41	7 71312·8	1413·6	33 88354·4	7605·3	149 44337	40317
·42	8 07019·9	1464·9	35 61383·5	7915·4	157 79115	42154
·43	8 44192·1	1517·1	37 42330·1	8237·0	166 56060	44061
·44	8 82881·8	1571·6	39 31515·8	8570·0	175 77081	46050
2·45	9 23143·4	1627·5	41 29273·6	8914·3	185 44166	48114
·46	9 65032·8	1685·2	43 35948·0	9271·2	195 59381	50266
·47	10 08607·7	1744·5	45 51895·9	9640·7	206 24877	52497
·48	10 53927·4	1805·6	47 77486·8	10022·8	217 42887	54822
·49	11 01053·1	1868·9	50 13102·9	10418·1	229 15735	57233
2·50	11 50048·0	1933·6	52 59139·7	10827·5	241 45834	59742

* These differences have been modified; see page A 4.

x	$P_{10}(x)$	$*\delta^2$	$P_{11}(x)$	$*\delta^2$	$10^{-1} . P_{12}(x)$	$*\delta^2$
2·50	11 50048·0	1933·6	52 59140	10827	241 4583	5974
·51	12 00976·9	2000·7	55 16007	11250	254 3569	6235
·52	12 53906·8	2069·4	57 84127	11688	267 8792	6505
·53	13 08906·5	2140·6	60 63938	12143	282 0522	6786
·54	13 66047·1	2213·3	63 55894	12609	296 9040	7079
2·55	14 25401·5	2288·9	66 60462	13094	312 4639	7382
·56	14 87045·1	2365·9	69 78127	13593	328 7622	7694
·57	15 51055·1	2445·7	73 09388	14111	345 8302	8024
·58	16 17511·2	2527·8	76 54763	14645	363 7008	8362
·59	16 86495·5	2611·9	80 14786	15198	382 4078	8712
2·60	17 58092·2	2699·0	83 90010	15766	401 9863	9079
·61	18 32388·3	2787·8	87 81004	16358	422 4729	9455
·62	19 09472·8	2880·0	91 88359	16965	443 9053	9848
·63	19 89437·7	2974·3	96 12683	17593	466 3228	1 0257
·64	20 72377·4	3071·5	100 54604	18242	489 7662	1 0676
2·65	21 58389·1	3171·2	105 14771	18910	514 2775	1 1114
·66	22 47572·5	3273·7	109 93853	19606	539 9005	1 1570
·67	23 40030·2	3379·7	114 92544	20317	566 6807	1 2035
·68	24 35868·0	3487·7	120 11556	21052	594 6648	1 2524
·69	25 35194·1	3599·4	125 51625	21815	623 9016	1 3028
2·70	26 38120·1	3713·8	131 13513	22596	654 4415	1 3549
·71	27 44760·5	3831·6	136 98002	23407	686 3367	1 4092
·72	28 55233·1	3952·4	143 05902	24240	719 6414	1 4650
·73	29 69658·7	4076·7	149 38047	25101	754 4115	1 5229
·74	30 88161·6	4204·1	155 95298	25988	790 7049	1 5833
2·75	32 10869	4337	162 78542	26903	828 5819	1 6450
·76	33 37913	4469	169 88694	27846	868 1044	1 7095
·77	34 69427	4608	177 26697	28817	909 3368	1 7761
·78	36 05550	4750	184 93523	29819	952 3457	1 8447
·79	37 46424	4895	192 90174	30852	997 1998	1 9162
2·80	38 92194	5044	201 17683	31916	1043 9705	1 9899
·81	40 43009	5200	209 77114	33013	1092 7315	2 0659
·82	41 99024	5356	218 69564	34142	1143 5589	2 1447
·83	43 60396	5517	227 96162	35306	1196 5315	2 2263
·84	45 27286	5685	237 58072	36503	1251 7309	2 3105
2·85	46 99861	5853	247 56492	37737	1309 2413	2 3976
·86	48 78290	6028	257 92656	39008	1369 1498	2 4874
·87	50 62748	6208	268 67835	40317	1431 5463	2 5807
·88	52 53415	6391	279 83338	41664	1496 5240	2 6766
·89	54 50474	6579	291 40512	43051	1564 1789	2 7759
2·90	56 54113	6772	303 40744	44476	1634 6103	2 8786
·91	58 64525	6973	315 85460	45948	1707 9209	2 9844
·92	60 81910	7172	328 76131	47458	1784 2166	3 0939
·93	63 06469	7384	342 14268	49011	1863 6069	3 2071
·94	65 38412	7595	356 01425	50614	1946 2049	3 3234
2·95	67 77951	7815	370 39204	52259	2032 1271	3 4442
·96	70 25306	8039	385 29251	53954	2121 4942	3 5684
·97	72 80701	8269	400 73260	55692	2214 4305	3 6970
·98	75 44366	8504	416 72971	57486	2311 0645	3 8293
2·99	78 16536	8746	433 30177	59326	2411 5286	3 9662
3·00	80 97453	8993	450 46719	61220	2515 9597	4 1072

* These differences have been modified; see page A 4.

5-2

x	$P_{10}(x)$	$*\delta^2$	$10^{-1} . P_{11}(x)$	δ^2	$10^{-2} . P_{12}(x)$	$*\delta^2$
3·00	80 97453	8993	450 4672	6123	2515 960	4 107
·01	83 87364	9247	468 2449	6318	2624 499	4 252
·02	86 86523	9506	486 6544	6518	2737 291	4 405
·03	89 95189	9773	505 7157	6724	2854 488	4 557
·04	93 13629	10042	525 4494	6934	2976 243	4 718
3·05	96 42113	10324	545 8765	7153	3102 717	4 882
·06	99 80922	10609	567 0189	7376	3234 074	5 054
·07	103 30341	10902	588 8989	7606	3370 485	5 225
·08	106 90663	11201	611 5395	7840	3512 123	5 409
·09	110 62187	11506	634 9641	8083	3659 171	5 595
3·10	114 45219	11822	659 1970	8333	3811 815	5 784
·11	118 40074	12141	684 2632	8587	3970 245	5 984
·12	122 47072	12472	710 1881	8850	4134 660	6 188
·13	126 66543	12809	736 9980	9120	4305 264	6 396
·14	130 98824	13151	764 7199	9395	4482 266	6 615
3·15	135 44258	13505	793 3813	9682	4665 884	6 837
·16	140 03198	13866	823 0109	9971	4856 340	7 066
·17	144 76005	14235	853 6376	1 0272	5053 863	7 303
·18	149 63048	14611	885 2915	1 0579	5258 690	7 547
·19	154 64704	14999	918 0033	1 0894	5471 065	7 798
3·20	159 81360	15393	951 8045	1 1219	5691 239	8 053
·21	165 13411	15797	986 7276	1 1550	5919 468	8 324
·22	170 61261	16210	1022 8057	1 1891	6156 021	8 593
·23	176 25323	16633	1060 0729	1 2242	6401 169	8 878
·24	182 06020	17065	1098 5643	1 2600	6655 196	9 165
3·25	188 03784	17506	1138 3157	1 2968	6918 390	9 466
·26	194 19056	17961	1179 3639	1 3346	7191 051	9 771
·27	200 52290	18421	1221 7467	1 3734	7473 485	10 087
·28	207 03947	18895	1265 5029	1 4129	7766 008	10 415
·29	213 74500	19374	1310 6720	1 4539	8068 947	10 746
3·30	220 64430	19872	1357 2950	1 4956	8382 634	11 091
·31	227 74233	20372	1405 4136	1 5384	8707 414	11 447
·32	235 04411	20892	1455 0706	1 5822	9043 642	11 809
·33	242 55482	21415	1506 3098	1 6274	9391 681	12 183
·34	250 27971	21955	1559 1764	1 6735	9751 905	12 568
3·35	258 22417	22506	1613 7165	1 7208	10124 699	12 965
·36	266 39371	23066	1669 9774	1 7693	10510 460	13 371
·37	274 79394	23642	1728 0076	1 8189	10909 594	13 790
·38	283 43061	24229	1787 8567	1 8699	11322 520	14 218
·39	292 30959	24826	1849 5757	1 9220	11749 667	14 662
3·40	301 43686	25440	1913 2167	1 9756	12191 478	15 117
·41	310 81855	26066	1978 8333	2 0302	12648 408	15 580
·42	320 46092	26703	2046 4801	2 0865	13120 921	16 065
·43	330 37035	27355	2116 2134	2 1439	13609 500	16 554
·44	340 55336	28023	2188 0906	2 2027	14114 636	17 063
3·45	351 01662	28703	2262 1705	2 2630	14636 837	17 581
·46	361 76693	29395	2338 5134	2 3249	15176 622	18 116
·47	372 81122	30107	2417 1812	2 3881	15734 526	18 665
·48	384 15660	30829	2498 2371	2 4529	16311 098	19 229
·49	395 81030	31568	2581 7459	2 5190	16906 902	19 808
3·50	407 77971	32323	2667 7737	2 5872	17522 517	20 403

* These differences have been modified; see page A 4.

x	$10^{-1}.P_{10}(x)$	δ^2	$10^{-2}.P_{11}(x)$	δ^2	$10^{-3}.P_{12}(x)$	δ^2
3.50	407 7797	3233	2667 774	2 587	17522 52	20 40
.51	420 0724	3309	2756 389	2 656	18158 54	21 02
.52	432 6960	3389	2847 660	2 728	18815 58	21 64
.53	445 6585	3467	2941 659	2 801	19494 26	22 29
.54	458 9677	3552	3038 459	2 875	20195 23	22 94
3.55	472 6321	3633	3138 134	2 952	20919 14	23 64
.56	486 6598	3719	3240 761	3 029	21666 69	24 31
.57	501 0594	3806	3346 417	3 109	22438 55	25 05
.58	515 8396	3895	3455 182	3 192	23235 46	25 76
.59	531 0093	3985	3567 139	3 275	24058 13	26 53
3.60	546 5775	4078	3682 371	3 360	24907 33	27 29
.61	562 5535	4173	3800 963	3 449	25783 82	28 10
.62	578 9468	4267	3923 004	3 537	26688 41	28 89
.63	595 7668	4367	4048 582	3 630	27621 89	29 74
.64	613 0235	4466	4177 790	3 723	28585 11	30 59
3.65	630 7268	4568	4310 721	3 819	29578 92	31 47
.66	648 8869	4673	4447 471	3 917	30604 20	32 36
.67	667 5143	4778	4588 138	4 018	31661 84	33 30
.68	686 6195	4887	4732 823	4 121	32752 78	34 24
.69	706 2134	4997	4881 629	4 224	33877 96	35 20
3.70	726 3070	5109	5034 659	4 332	35038 34	36 21
.71	746 9115	5225	5192 021	4 443	36234 93	37 22
.72	768 0385	5340	5353 826	4 553	37468 74	38 28
.73	789 6995	5461	5520 184	4 670	38740 83	39 33
.74	811 9066	5582	5691 212	4 785	40052 25	40 45
3.75	834 6719	5705	5867 025	4 905	41404 12	41 57
.76	858 0077	5832	6047 743	5 029	42797 56	42 73
.77	881 9267	5961	6233 490	5 152	44233 73	43 90
.78	906 4418	6092	6424 389	5 282	45713 80	45 12
.79	931 5661	6225	6620 570	5 411	47238 99	46 36
3.80	957 3129	6361	6822 162	5 545	48810 54	47 65
.81	983 6958	6501	7029 299	5 682	50429 74	48 93
.82	1010 7288	6642	7242 118	5 821	52097 87	50 29
.83	1038 4260	6785	7460 758	5 963	53816 29	51 64
.84	1066 8017	6934	7685 361	6 108	55586 35	53 06
3.85	1095 8708	7081	7916 072	6 258	57409 47	54 48
.86	1125 6480	7235	8153 041	6 409	59287 07	55 97
.87	1156 1487	7391	8396 419	6 564	61220 64	57 46
.88	1187 3885	7547	8646 361	6 723	63211 67	59 01
.89	1219 3830	7709	8903 026	6 884	65261 71	60 60
3.90	1252 1484	7874	9166 575	7 050	67372 35	62 21
.91	1285 7012	8040	9437 174	7 218	69545 20	63 86
.92	1320 0580	8210	9714 991	7 390	71781 91	65 57
.93	1355 2358	8384	10000 198	7 567	74084 19	67 31
.94	1391 2520	8561	10292 972	7 745	76453 78	69 07
3.95	1428 1243	8739	10593 491	7 930	78892 44	70 91
.96	1465 8705	8924	10901 940	8 116	81402 01	72 76
.97	1504 5091	9108	11218 505	8 308	83984 34	74 68
.98	1544 0585	9300	11543 378	8 502	86641 35	76 62
3.99	1584 5379	9492	11876 753	8 702	89374 98	78 62
4.00	1625 9665	9689	12218 830	8 905	92187 23	80 67

x	$10^{-1} \cdot P_{10}(x)$	δ^2	$10^{-2} \cdot P_{11}(x)$	δ^2	$10^{-3} \cdot P_{12}(x)$	δ^2
4·00	1625 9665	9689	12218 830	8 905	92187 23	80 67
·01	1668 3640	9889	12569 812	9 112	95080 15	82 75
·02	1711 7504	1 0093	12929 906	9 325	98055 82	84 91
·03	1756 1461	1 0300	13299 325	9 539	1 01116 40	87 09
·04	1801 5718	1 0511	13678 283	9 761	1 04264 07	89 32
4·05	1848 0486	1 0727	14067 002	9 986	1 07501 06	91 63
·06	1895 5981	1 0945	14465 707	10 215	1 10829 68	93 96
·07	1944 2421	1 1167	14874 627	10 449	1 14252 26	96 37
·08	1994 0028	1 1395	15293 996	10 688	1 17771 21	98 82
·09	2044 9030	1 1625	15724 053	10 933	1 21388 98	101 33
4·10	2096 9657	1 1859	16165 043	11 180	1 25108 08	103 89
·11	2150 2143	1 2098	16617 213	11 435	1 28931 07	106 52
·12	2204 6727	1 2341	17080 818	11 694	1 32860 58	109 20
·13	2260 3652	1 2589	17556 117	11 957	1 36899 29	111 95
·14	2317 3166	1 2839	18043 373	12 226	1 41049 95	114 76
4·15	2375 5519	1 3095	18542 855	12 501	1 45315 37	117 62
·16	2435 0967	1 3357	19054 838	12 782	1 49698 41	120 55
·17	2495 9772	1 3619	19579 603	13 067	1 54202 00	123 56
·18	2558 2196	1 3891	20117 435	13 358	1 58829 15	126 62
·19	2621 8511	1 4164	20668 625	13 655	1 63582 92	129 75
4·20	2686 8990	1 4442	21233 470	13 958	1 68466 44	132 96
·21	2753 3911	1 4725	21812 273	14 267	1 73482 92	136 24
·22	2821 3557	1 5015	22405 343	14 581	1 78635 64	139 57
·23	2890 8218	1 5308	23012 994	14 903	1 83927 93	143 00
·24	2961 8187	1 5605	23635 548	15 230	1 89363 22	146 49
4·25	3034 3761	1 5908	24273 332	15 563	1 94945 00	150 07
·26	3108 5243	1 6218	24926 679	15 903	2 00676 85	153 71
·27	3184 2943	1 6531	25595 929	16 250	2 06562 41	157 45
·28	3261 7174	1 6848	26281 429	16 604	2 12605 42	161 25
·29	3340 8253	1 7175	26983 533	16 962	2 18809 68	165 14
4·30	3421 6507	1 7503	27702 599	17 331	2 25179 08	169 12
·31	3504 2264	1 7838	28438 996	17 704	2 31717 60	173 19
·32	3588 5859	1 8179	29193 097	18 086	2 38429 31	177 33
·33	3674 7633	1 8526	29965 284	18 472	2 45318 35	181 57
·34	3762 7933	1 8877	30755 943	18 871	2 52388 96	185 91
4·35	3852 7110	1 9235	31565 473	19 272	2 59645 48	190 32
·36	3944 5522	1 9600	32394 275	19 683	2 67092 32	194 84
·37	4038 3534	1 9969	33242 760	20 103	2 74734 00	199 45
·38	4134 1515	2 0345	34111 348	20 528	2 82575 13	204 16
·39	4231 9841	2 0727	35000 464	20 964	2 90620 42	208 97
4·40	4331 8894	2 1115	35910 544	21 405	2 98874 68	213 89
·41	4433 9062	2 1511	36842 029	21 856	3 07342 83	218 90
·42	4538 0741	2 1910	37795 370	22 317	3 16029 88	224 01
·43	4644 4330	2 2319	38771 028	22 784	3 24940 94	229 25
·44	4753 0238	2 2732	39769 470	23 259	3 34081 25	234 58
4·45	4863 8878	2 3154	40791 171	23 747	3 43456 14	240 02
·46	4977 0672	2 3582	41836 619	24 239	3 53071 05	245 59
·47	5092 6048	2 4015	42906 306	24 743	3 62931 55	251 27
·48	5210 5439	2 4457	44000 736	25 257	3 73043 32	257 06
·49	5330 9287	2 4906	45120 423	25 777	3 83412 15	262 98
4·50	5453 8041	2 5362	46265 887	26 310	3 94043 96	269 00

x	$10^{-2} . P_{10}(x)$	δ^2	$10^{-3} . P_{11}(x)$	δ^2	$10^{-4} . P_{12}(x)$	δ^2
4·50	5453 804	2 537	46265 89	26 30	3 94044 0	269 0
·51	5579 216	2 582	47437 66	26 86	4 04944 8	275 1
·52	5707 210	2 629	48636 29	27 39	4 16120 7	281 6
·53	5837 833	2 678	49862 31	27 97	4 27578 2	287 8
·54	5971 134	2 725	51116 30	28 53	4 39323 5	294 4
4·55	6107 160	2 776	52398 82	29 12	4 51363 2	301 1
·56	6245 962	2 825	53710 46	29 71	4 63704 0	307 9
·57	6387 589	2 875	55051 81	30 31	4 76352 7	314 9
·58	6532 091	2 929	56423 47	30 92	4 89316 3	321 9
·59	6679 522	2 980	57826 05	31 55	5 02601 8	329 2
4·60	6829 933	3 033	59260 18	32 19	5 16216 5	336 6
·61	6983 377	3 087	60726 50	32 83	5 30167 8	344 1
·62	7139 908	3 143	62225 65	33 49	5 44463 2	351 8
·63	7299 582	3 199	63758 29	34 16	5 59110 4	359 6
·64	7462 455	3 254	65325 09	34 84	5 74117 2	367 6
4·65	7628 582	3 313	66926 73	35 54	5 89491 6	375 8
·66	7798 022	3 371	68563 91	36 24	6 05241 8	384 0
·67	7970 833	3 430	70237 33	36 96	6 21376 0	392 6
·68	8147 074	3 491	71947 71	37 70	6 37902 8	401 2
·69	8326 806	3 552	73695 79	38 44	6 54830 8	410 1
4·70	8510 090	3 613	75482 31	39 19	6 72168 9	418 9
·71	8696 987	3 677	77308 02	39 96	6 89925 9	428 2
·72	8887 561	3 741	79173 69	40 76	7 08111 1	437 5
·73	9081 876	3 806	81080 12	41 54	7 26733 8	447 1
·74	9279 997	3 872	83028 09	42 36	7 45803 6	456 7
4·75	9481 990	3 938	85018 42	43 19	7 65330 1	466 7
·76	9687 921	4 007	87051 94	44 02	7 85323 3	476 7
·77	9897 859	4 076	89129 48	44 87	8 05793 2	487 1
·78	10111 873	4 145	91251 89	45 76	8 26750 2	497 6
·79	10330 032	4 217	93420 06	46 62	8 48204 8	508 2
4·80	10552 408	4 289	95634 85	47 53	8 70167 6	519 1
·81	10779 073	4 361	97897 17	48 44	8 92649 5	530 3
·82	11010 099	4 437	1 00207 93	49 37	9 15661 7	541 5
·83	11245 562	4 512	1 02568 06	50 32	9 39215 4	553 2
·84	11485 537	4 588	1 04978 51	51 28	9 63322 3	564 8
4·85	11730 100	4 666	1 07440 24	52 26	9 87994 0	576 9
·86	11979 329	4 744	1 09954 23	53 24	10 13242 6	589 0
·87	12233 302	4 825	1 12521 46	54 27	10 39080 2	601 6
·88	12492 100	4 905	1 15142 96	55 28	10 65519 4	614 1
·89	12755 803	4 989	1 17819 74	56 34	10 92572 7	627 2
4·90	13024 495	5 071	1 20552 86	57 40	11 20253 2	640 2
·91	13298 258	5 155	1 23343 38	58 47	11 48573 9	653 7
·92	13577 176	5 243	1 26192 37	59 57	11 77548 3	667 3
·93	13861 337	5 330	1 29100 93	60 70	12 07190 0	681 3
·94	14150 828	5 416	1 32070 19	61 83	12 37513 0	695 5
4·95	14445 735	5 509	1 35101 28	62 97	12 68531 5	709 8
·96	14746 151	5 598	1 38195 34	64 16	13 00259 8	724 6
·97	15052 165	5 690	1 41353 56	65 35	13 32712 7	739 5
·98	15363 869	5 785	1 44577 13	66 55	13 65905 1	754 9
4·99	15681 358	5 880	1 47867 25	67 79	13 99852 4	770 4
5·00	16004 727	5 975	1 51225 16	69 04	14 34570 1	786 2

x	$10^{-2} \cdot P_{10}(x)$	δ^2	$10^{-3} \cdot P_{11}(x)$	δ^2	$10^{-5} \cdot P_{12}(x)$	δ^2
5·00	16004 727	5 975	1 51225 16	69 04	14 34570	786
·01	16334 071	6 074	1 54652 11	70 32	14 70074	802
·02	16669 489	6 173	1 58149 38	71 60	15 06380	819
·03	17011 080	6 273	1 61718 25	72 93	15 43505	836
·04	17358 944	6 376	1 65360 05	74 25	15 81466	852
5·05	17713 184	6 478	1 69076 10	75 61	16 20279	869
·06	18073 902	6 583	1 72867 76	77 00	16 59961	889
·07	18441 203	6 691	1 76736 42	78 39	17 00532	904
·08	18815 195	6 797	1 80683 47	79 83	17 42007	925
·09	19195 984	6 908	1 84710 35	81 25	17 84407	942
5·10	19583 681	7 018	1 88818 48	82 75	18 27749	961
·11	19978 396	7 130	1 93009 36	84 22	18 72052	981
·12	20380 241	7 245	1 97284 46	85 75	19 17336	1001
·13	20789 331	7 360	2 01645 31	87 29	19 63621	1019
·14	21205 781	7 477	2 06093 45	88 86	20 10925	1042
5·15	21629 708	7 597	2 10630 45	90 44	20 59271	1061
·16	22061 232	7 717	2 15257 89	92 07	21 08678	1082
·17	22500 473	7 839	2 19977 40	93 70	21 59167	1105
·18	22947 553	7 962	2 24790 61	95 38	22 10761	1125
·19	23402 595	8 089	2 29699 20	97 06	22 63480	1149
5·20	23865 726	8 215	2 34704 85	98 80	23 17348	1171
·21	24337 072	8 345	2 39809 30	100 53	23 72387	1193
·22	24816 763	8 476	2 45014 28	102 31	24 28619	1218
·23	25304 930	8 608	2 50321 57	104 12	24 86069	1241
·24	25801 705	8 742	2 55732 98	105 94	25 44760	1266
5·25	26307 222	8 879	2 61250 33	107 82	26 04717	1289
·26	26821 618	9 017	2 66875 50	109 69	26 65963	1316
·27	27345 031	9 157	2 72610 36	111 61	27 28525	1342
·28	27877 601	9 298	2 78456 83	113 56	27 92429	1366
·29	28419 469	9 442	2 84416 86	115 54	28 57699	1393
5·30	28970 779	9 588	2 90492 43	117 55	29 24362	1422
·31	29531 677	9 736	2 96685 55	119 59	29 92447	1446
·32	30102 311	9 885	3 02998 26	121 65	30 61978	1477
·33	30682 830	10 037	3 09432 62	123 77	31 32986	1504
·34	31273 386	10 191	3 15990 75	125 89	32 05498	1533
5·35	31874 133	10 346	3 22674 77	128 06	32 79543	1562
·36	32485 226	10 504	3 29486 85	130 26	33 55150	1593
·37	33106 823	10 664	3 36429 19	132 50	34 32350	1622
·38	33739 084	10 825	3 43504 03	134 76	35 11172	1653
·39	34382 170	10 991	3 50713 63	137 06	35 91647	1685
5·40	35036 247	11 157	3 58060 29	139 41	36 73807	1718
·41	35701 481	11 324	3 65546 36	141 77	37 57685	1748
·42	36378 039	11 496	3 73174 20	144 18	38 43311	1784
·43	37066 093	11 668	3 80946 22	146 63	39 30721	1815
·44	37765 815	11 844	3 88864 87	149 11	40 19946	1850
5·45	38477 381	12 022	3 96932 63	151 62	41 11021	1886
·46	39200 969	12 201	4 05152 01	154 18	42 03982	1919
·47	39936 758	12 383	4 13525 57	156 77	42 98862	1957
·48	40684 930	12 569	4 22055 90	159 41	43 95699	1993
·49	41445 671	12 754	4 30745 64	162 07	44 94529	2030
5·50	42219 166	12 945	4 39597 45	164 80	45 95389	2067

x	$10^{-3} . P_{10}(x)$	δ^2	$10^{-4} . P_{11}(x)$	δ^2	$10^{-5} . P_{12}(x)$	δ^2
5·50	42219 17	12 94	4 39597 5	164 7	45 95389	2067
·51	43005 61	13 13	4 48614 1	167 5	46 98316	2107
·52	43805 18	13 34	4 57798 2	170 4	48 03350	2145
·53	44618 09	13 52	4 67152 7	173 1	49 10529	2185
·54	45444 52	13 73	4 76680 3	176 1	50 19893	2225
5·55	46284 68	13 93	4 86384 0	178 9	51 31482	2267
·56	47138 77	14 14	4 96266 6	182 0	52 45338	2307
·57	48007 00	14 33	5 06331 2	184 9	53 61501	2351
·58	48889 56	14 55	5 16580 7	187 9	54 80015	2393
·59	49786 67	14 77	5 27018 1	191 1	56 00922	2438
5·60	50698 55	14 97	5 37646 6	194 3	57 24267	2481
·61	51625 40	15 20	5 48469 4	197 2	58 50093	2528
·62	52567 45	15 42	5 59489 4	200 7	59 78447	2572
·63	53524 92	15 63	5 70710 1	203 9	61 09373	2621
·64	54498 02	15 88	5 82134 7	207 2	62 42920	2666
5·65	55487 00	16 08	5 93766 5	210 6	63 79133	2717
·66	56492 06	16 34	6 05608 9	213 9	65 18063	2763
·67	57513 46	16 55	6 17665 2	217 6	66 59756	2816
·68	58551 41	16 81	6 29939 1	220 9	68 04265	2864
·69	59606 17	17 04	6 42433 9	224 6	69 51638	2917
5·70	60677 97	17 28	6 55153 3	228 1	71 01928	2969
·71	61767 05	17 54	6 68100 8	231 9	72 55187	3022
·72	62873 67	17 77	6 81280 2	235 6	74 11468	3076
·73	63998 06	18 04	6 94695 2	239 3	75 70825	3131
·74	65140 49	18 30	7 08349 5	243 2	77 33313	3186
5·75	66301 22	18 54	7 22247 0	247 0	78 98987	3244
·76	67480 49	18 81	7 36391 5	251 1	80 67905	3301
·77	68678 57	19 09	7 50787 1	254 9	82 40124	3359
·78	69895 74	19 34	7 65437 6	259 1	84 15702	3419
·79	71132 25	19 63	7 80347 2	263 2	85 94699	3470
5·80	72388 39	19 89	7 95520 0	267 3	87 77175	3540
·81	73664 42	20 17	8 10960 1	271 5	89 63191	3602
·82	74960 62	20 47	8 26671 7	275 9	91 52809	3667
·83	76277 29	20 74	8 42659 2	280 2	93 46094	3729
·84	77614 70	21 03	8 58926 9	284 5	95 43108	3796
5·85	78973 14	21 33	8 75479 1	289 2	97 43918	3861
·86	80352 91	21 62	8 92320 5	293 4	99 48589	3929
·87	81754 30	21 93	9 09455 3	298 3	101 57189	3997
·88	83177 62	22 22	9 26888 4	302 8	103 69786	4067
·89	84623 16	22 54	9 44624 3	307 5	105 86450	4136
5·90	86091 24	22 85	9 62667 7	312 4	108 07250	4210
·91	87582 17	23 16	9 81023 5	317 1	110 32260	4281
·92	89096 26	23 47	9 99696 4	322 2	112 61551	4355
·93	90633 82	23 81	10 18691 5	327 0	114 95197	4430
·94	92195 19	24 13	10 38013 6	332 2	117 33273	4506
5·95	93780 69	24 46	10 57667 9	337 2	119 75855	4584
·96	95390 65	24 80	10 77659 4	342 5	122 23021	4662
·97	97025 41	25 12	10 97993 4	347 7	124 74849	4742
·98	98685 29	25 49	11 18675 1	353 0	127 31419	4823
5·99	1 00370 66	25 81	11 39709 8	358 5	129 92812	4905
6·00	1 02081 84	26 17	11 61103 0	364 0	132 59110	4988

x	$P_2(x)$†	$P_3(x)$	δ^2	$P_4(x)$	δ^2	$P_5(x)$	*δ^2	$P_6(x)$	*δ^2
6·0	53·500	531·0000	9000	5535·375	18·826	59357·25	336·98	6 48316·9	5526·6
6·1	55·315	558·3025	9150	5918·393	19·461	64537·31	354·25	7 16810·3	5907·1
6·2	57·160	586·5200	9300	6320·872	20·107	70071·72	372·04	7 91214·4	6307·5
6·3	59·035	615·6675	9450	6743·458	20·763	75978·28	390·46	8 71929·6	6727·2
6·4	60·940	645·7600	9600	7186·807	21·430	82275·41	409·46	9 59375·8	7167·9
6·5	62·875	676·8125	9750	7651·586	22·107	88982·11	429·06	10 53993·8	7629·5
6·6	64·840	708·8400	9900	8138·472	22·795	96117·98	449·27	11 56245·4	8113·7
6·7	66·835	741·8575	1·0050	8648·153	23·493	1 03703·24	470·13	12 66614·7	8619·4
6·8	68·860	775·8800	1·0200	9181·327	24·202	1 11758·74	491·59	13 85607·8	9149·5
6·9	70·915	810·9225	1·0350	9738·703	24·921	1 20305·95	513·72	15 13754·7	9702·8
7·0	73·000	847·0000	1·0500	10321·000	25·651	1 29367·00	536·47	16 51609·0	10281·1
7·1	75·115	884·1275	1·0650	10928·948	26·391	1 38964·65	559·92	17 99749·1	10884·9
7·2	77·260	922·3200	1·0800	11563·287	27·142	1 49122·34	584·03	19 58778·9	11514·6
7·3	79·435	961·5925	1·0950	12224·768	27·903	1 59864·18	608·79	21 29328·3	12171·4
7·4	81·640	1001·9600	1·1100	12914·152	28·675	1 71214·94	634·27	23 12054·2	12855·9
7·5	83·875	1043·4375	1·1250	13632·211	29·457	1 83200·10	660·44	25 07641·2	13568·6
7·6	86·140	1086·0400	1·1400	14379·727	30·250	1 95845·83	687·34	27 16802·2	14310·5
7·7	88·435	1129·7825	1·1550	15157·493	31·053	2 09179·03	714·91	29 40279·3	15082·8
7·8	90·760	1174·6800	1·1700	15966·312	31·867	2 23227·28	743·24	31 78844·8	15885·2
7·9	93·115	1220·7475	1·1850	16806·998	32·691	2 38018·91	772·33	34 33301·4	16719·8
8·0	95·500	1268·0000	1·2000	17680·375	33·526	2 53583·00	802·12	37 04483·7	17586·4
8·1	97·915	1316·4525	1·2150	18587·278	34·371	2 69949·35	832·69	39 93258·5	18486·1
8·2	100·360	1366·1200	1·2300	19528·552	35·227	2 87148·53	864·03	43 00525·8	19420·3
8·3	102·835	1417·0175	1·2450	20505·053	36·093	3 05211·88	896·11	46 27219·8	20389·3
8·4	105·340	1469·1600	1·2600	21517·647	36·970	3 24171·49	929·04	49 74309·6	21393·9
8·5	107·875	1522·5625	1·2750	22567·211	37·857	3 44060·3	962·6	53 42800	22435
8·6	110·440	1577·2400	1·2900	23654·632	38·755	3 64911·9	997·3	57 33732	23514
8·7	113·035	1633·2075	1·3050	24780·808	39·663	3 86760·9	1032·4	61 48185	24632
8·8	115·660	1690·4800	1·3200	25946·647	40·582	4 09642·5	1068·7	65 87277	25786
8·9	118·315	1749·0725	1·3350	27153·068	41·511	4 33592·9	1105·5	70 52163	26985
9·0	121·000	1809·0000	1·3500	28401·000	42·451	4 58649·0	1143·3	75 44041	28221
9·1	123·715	1870·2775	1·3650	29691·383	43·401	4 84848·6	1182·1	80 64148	29504
9·2	126·460	1932·9200	1·3800	31025·167	44·362	5 12230·4	1221·6	86 13766	30822
9·3	129·235	1996·9425	1·3950	32403·313	45·333	5 40833·9	1261·8	91 94215	32194
9·4	132·040	2062·3600	1·4100	33826·792	46·315	5 70699·4	1303·3	98 06865	33601
9·5	134·875	2129·1875	1·4250	35296·586	47·307	6 01868·3	1345·2	104 53125	35062
9·6	137·740	2197·4400	1·4400	36813·687	48·310	6 34382·6	1388·2	111 34455	36564
9·7	140·635	2267·1325	1·4550	38379·098	49·323	6 68285·3	1432·5	118 52358	38119
9·8	143·560	2338·2800	1·4700	39993·832	50·347	7 03620·6	1476·9	126 08388	39717
9·9	146·515	2410·8975	1·4850	41658·913	51·381	7 40433·1	1523·1	134 04145	41370
10·0	149·500	2485·0000	1·5000	43375·375	52·426	7 78768·8	1569·4	142 41281	43070
10·1	152·515	2560·6025	1·5150	45144·263	53·481	8 18674·2	1617·5	151 21497	44827
10·2	155·560	2637·7200	1·5300	46966·632	54·547	8 60197·2	1665·9	160 46549	46629
10·3	158·635	2716·3675	1·5450	48843·548	55·623	9 03386·3	1715·5	170 18241	48495
10·4	161·740	2796·5600	1·5600	50776·087	56·710	9 48291·1	1766·1	180 38437	50407
10·5	164·875	2878·3125	1·5750	52765·336	57·807	9 94962·2	1817·6	191 09051	52382
10·6	168·040	2961·6400	1·5900	54812·392	58·915	10 43451·1	1870·2	202 32057	54409
10·7	171·235	3046·5575	1·6050	56918·363	60·033	10 93810·4	1923·7	214 09483	56497
10·8	174·460	3133·0800	1·6200	59084·367	61·162	11 46093·6	1978·3	226 43417	58645
10·9	177·715	3221·2225	1·6350	61311·533	62·301	12 00355·3	2033·8	239 36007	60853
11·0	181·000	3311·0000	1·6500	63601·000	63·451	12 56651·0	2090·4	252 89461	63121

* These differences have been modified; see page A 4. † For $P_2(x)$, δ^2 is $+30$ throughout.

For EU product safety concerns, contact us at Calle de José Abascal, 56–1°,
28003 Madrid, Spain or eugpsr@cambridge.org.

www.ingramcontent.com/pod-product-compliance
Ingram Content Group UK Ltd.
Pitfield, Milton Keynes, MK11 3LW, UK
UKHW051305151225
465965UK00022B/450